44

derangements

and the shape of persistence

Thad Roberts

44 derangements and the shape of persistence

By Thad Roberts

First print date 08-08-2021

This book is the extended version of Source Code, first print date 03-18-2021

Published in the United States by Thad Roberts

ISBN 978-0-9963942-9-1

Cover by Jeff Chapple
Figures by Thad Roberts and Jeff Chapple
Original hyperbolic figure eight knot STL by Henry Segerman
https://www.thingiverse.com/thing:1668611

Other books by Thad:

Einstein's Intuition: Visualizing Nature in Eleven Dimensions

Moon Rock: Mare Crisium

Passages

A Perfect Universe

Source Code: the balance of persistence

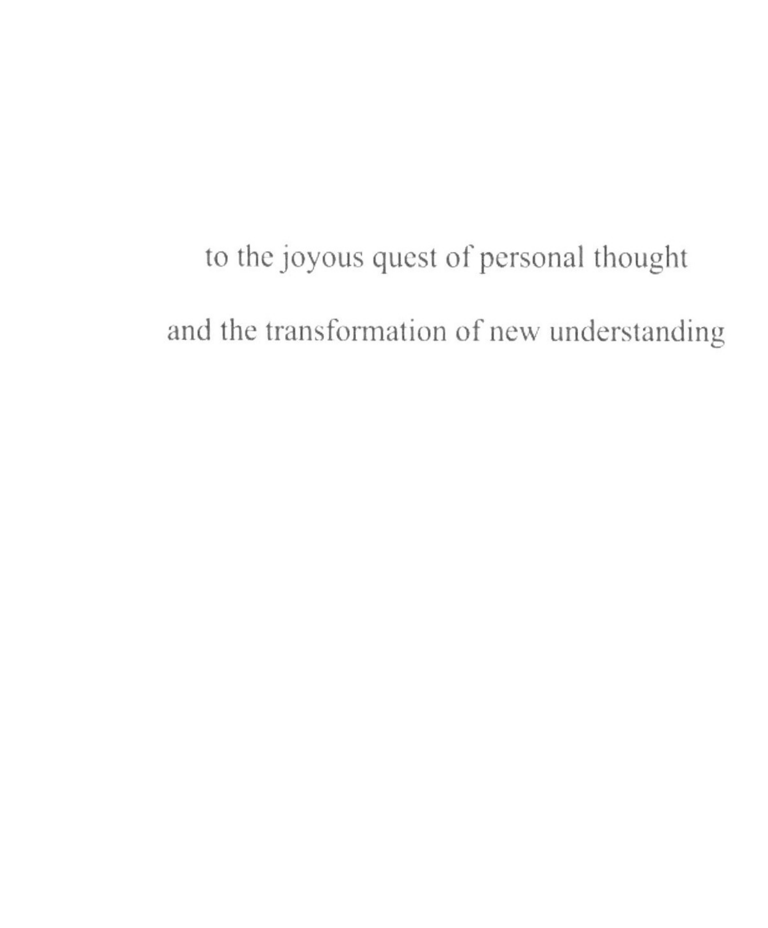

to the joyous quest of personal thought

and the transformation of new understanding

This book composes the most accurate, complete, and ontologically empowering story of physical reality that's ever been told. In short, in this book it is discovered that:

1. The balance of the hyperbolic figure eight knot decomposes into 5 actions with Planck constant boundaries, defining a single domain under 5 part persistent action whose limits define the boundaries of time, space, charge, mass and temperature.

2. The external facing boundaries of that balance connect via the hyperbolic vortex equation, partitioning the Planck *charge* and Planck *mass* boundaries into the exact charge and mass values that define the fundamental particles of matter.

3. Every unique boundary intersection within this minimum balance defines a constant of Nature.

4. The Planck union is dually hyperbolic, structurally determined by the ideal hyperbolic connection (defined by e, the base of natural logarithms) *and* the ideal hyperbolical partition function (the gamma function). And,

5. The Riemann zeta function, the octonions, and calculus are defined by the combinatorics of this minimum partition balance.

In summary, in this book we find that the simplest self-balanced geometry defines the minimal persistent arena, and the division parameters of that arena define the fundamental parameters of physical reality.

Preface: the search

> *"The task is not to see what has never been seen before, but to think what has never been thought before about what you see every day."*
>
> *Erwin Schrödinger*

In an early attempt to construct a story that accounts for reality's persistent physical properties, Plato advanced the idea that the elemental building blocks of geometry are also the elemental building blocks of reality; that the 5 elemental shapes (constructible from faces with equal length sides, the Platonic solids) are *atoms* of fire, air, earth, water, and aether—the primordial substance everything else was contained in. The reach of this idea triggered a categorical awakening, transforming what it meant to be conscious.

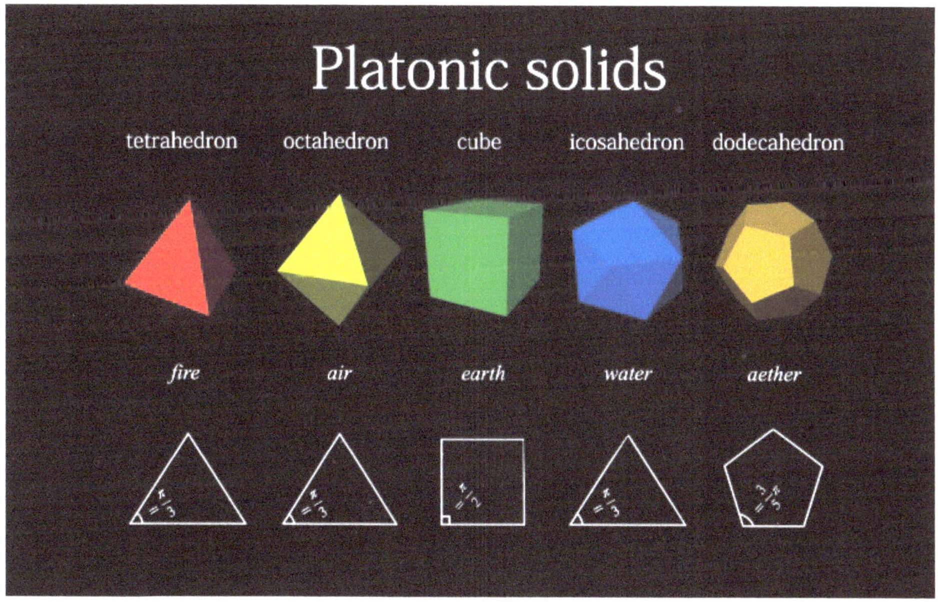

The 5 Platonic solids

Waking up to the idea that geometry plays a role in the construction of reality makes it possible for us to walk *ourselves* out of the cave of ignorance. A *geometry* is something we can say more about, something we can investigate, measure, and define the internal logic of. The possession of this knowledge transforms its holder into someone capable of seeking greater clarity and finding it, someone capable of self-transformation.

For the first couple of eons after Plato's geometric insight lit the caverns of consciousness, humanity's search for Nature's implicitly obtainable logic, or answering the call of existence and making one's best attempt to make sense of things, meant studying the Platonic solids. If you wanted to uncover the secrets of reality, you would spend your hours staring at these elemental shapes, trying to divine anything else they might have to say.

After 2000 years of intense focus Leonard Euler came along and, with little more than a glance, noticed that all of the Platonic solids are connected by a single elementary geometric relationship. Every one of them maintains the same balance of zero, one, and two-dimensional features (vertices, edges, faces). That is, each shape's number of vertices, minus its number of edges, plus its number of faces is always equal to 2.

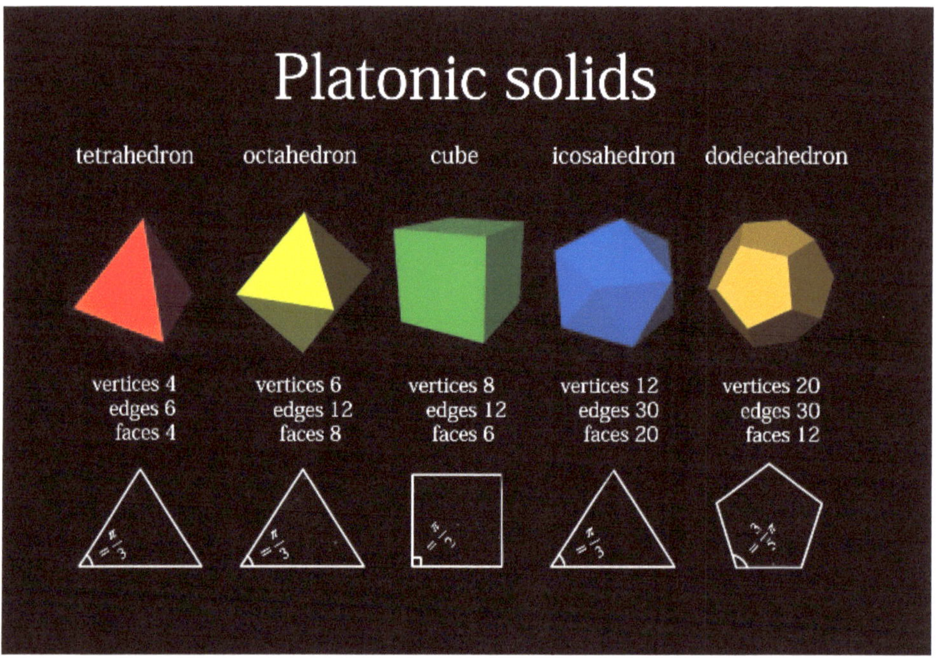

Euler's surprise relationship. Until Euler came along the number of edges these objects had were "hidden features", missed by everyone.

$$\text{vertices} - \text{edges} + \text{faces} = 2$$

$$V - E + F = 2$$

However obvious and simple this fact may now appear, the question that we must address is, "How did everyone else miss it?" Despite the intense interest in these geometric forms, and the generations of people searching them for further clues, nobody had ever found what Euler found. Why? It doesn't take much time to count up the properties of these shapes and compare them. Yet nobody had. Why not?

The answer is that, before Euler, nobody had imagined that an object's *number of sides* was a feature worthy of attention. They had never seen anyone else care about such a feature, so it just hadn't occurred. Euler was the first to count up these features and compare them because he was the first to imagine the option. He woke us up to that concept (and *many* more), lifting the ceiling of what it means *to think* and, therefore, *to be*.

Conscious thinkers have only two routes to new thought. Through example, or through genuine new reaches of imagination. Just as it is usually easier to verify an answer than to find one, it is usually easier to follow a line of logic laid out for you, than to lay one out for yourself. Which is why conscious thinkers almost always follow.

The trouble is, that if we are always just following, then we can get seriously stuck. If nobody has ever shown us an example of being interested in the number of edges an object has, we can all stare at something for 2000 years without seeing what's right in front of our eyes.

Euler looked into the world with new eyes, under his own investigation, instead of looking as he had been directed to. And, as a consequence, the world inherited new conceptual powers.

Another famous example of how difficult it can be to take the next logical next step without anyone showing us how, comes from our history of conceptualizing gravity. For millennia humans personally pondered *why* things fall to the ground, but it wasn't until Galileo came along that anyone thought to measure *how* things fall under the influence of gravity.

Galileo's measurements could be performed by nearly any human within minutes, with no cost (like rolling a ball down an inclined plane and marking the ball's position at successive increments of time and then comparing the distance between those marks). But before Galileo nobody had tried. Nobody had *thought* to.

By extending the powers of *measurement* to gravity, Galileo became the first to see it more clearly; famously finding that gravity's action depends on an inverse square law. In other words, he discovered that *gravity* is *geometric*.

This fact about gravity had always been true, it had always been staring us in the face, screaming from every freefall occurrence ever witnessed, but nobody had seen it before. Because nobody thought of gravity as a thing to physically measure, a thing to *look into* that way.

Through his efforts Galileo came to one of the greatest insights of all time, that the truth about reality is right in front of our eyes. The task is to see it more clearly, to more explicitly conceive its true form.

> *"Philosophy is written in this grand book, the universe, which stands continually open to our gaze. But the book cannot be understood unless one first learns to comprehend the language and read the letters in which it is composed. It is written in the language of mathematics, and its characters are triangles, circles, and other geometric figures without which it is humanly impossible to understand a single word of it; without these, one wanders about in a dark labyrinth."*
>
> *Galileo Galilei*

Today's best geometric description of the fundamental structure of reality is called *quantum field theory*. Correctly parametrized, this construction produces *all* of the actions of quantum mechanics and special relativity, a feat that earns it the title of being the pinnacle achievement of science.

The problem is that, although quantum field theory makes the right predictions, nobody understands why it has that particular construction. That is, the parameters of quantum field theory have no story themselves. Every single one of them remains utterly unexplained, known from experimental measurement only.

In short, we cannot explain *why* quantum field theory is constructed as it is, we cannot predict its parameters to any degree of accuracy at all, but once we construct our field theory with those particular parameters the actions of quantum mechanics and special relativity are reproduced in full.

> *"Look into Nature, and then you will understand it better."*
>
> *Albert Einstein*

Today, any serious investigation of reality involves staring at the parameters of quantum field theory, in search of the logic connecting them. That is, the modern version of the quest to tell the accurate story of reality's construction (now called *the theory of everything*) literally boils down to explaining the parameters of quantum field theory, explaining why they are what they are.

the parameters of quantum field theory

$m_e = 9.1093837015(28) \times 10^{-31}\ kg$ electron mass
$m_+ = 1.67262192369(51) \times 10^{-27}\ kg$ proton mass
$m_N = 1.67492749804(95) \times 10^{-27}\ kg$ neutron mass
$m_c = 2.272(63) \times 10^{-27}\ kg$ charm quark mass
$m_d = 8.3(07) \times 10^{-30}\ kg$ down quark mass
$m_u = 3.92(89) \times 10^{-30}\ kg$ up quark mass
$m_s = 1.69(16) \times 10^{-28}\ kg$ strange quark mass
$m_b = 7.45(07) \times 10^{-27}\ kg$ beauty (bottom) quark mass
$m_t = 3.084(07) \times 10^{-25}\ kg$ truth (top) quark mass
$m_H = 2.2315(28) \times 10^{-25}\ kg$ Higgs boson mass
$m_Z = 1.625566(38) \times 10^{-25}\ kg$ Z boson mass
$m_W = 1.43288(21) \times 10^{-25}\ kg$ W boson mass
$m_\tau = 3.16754(21) \times 10^{-27}\ kg$ tau mass
$m_\mu = 1.883531627(42) \times 10^{-28}\ kg$ muon mass
$m_{\nu_\tau} => 0\ kg\ ???$ tau neutrino mass
$m_{\nu_\mu} => 0\ kg\ ???$ muon neutrino mass
$m_{\nu_e} => 0\ kg\ ???$ electron neutrino mass

$\alpha = 7.2973525698(24) \times 10^{-3}$ fine-structure constant
$e = 1.602176565(35) \times 10^{-19}\ C$ electron charge
$\lambda_C = 2.4263102389(16) \times 10^{-12}\ m$ Compton wavelength
$K_J = 4.835978484 \times 10^{14}\ sC/m^2kg$ Josephson constant
$\hbar = 1.054571726(47) \times 10^{-34}\ m^2kg/s$ Planck's constant
$\varepsilon_0 = 8.8541878128(13) \times 10^{-12}\ s^2C^2/m^3kg$ electric constant
$\kappa = 8.9875517923(14) \times 10^{9}\ m^3kg/s^2C^2$ Coulomb's constant
$H_C = 3.87404614(17) \times 10^{-5}\ C^2/m^2kg$ quantized Hall conductance
$\mu_B = 9.274009994(57) \times 10^{-24}\ m^2C/s$ Bohr magneton
$\mu_0 = 1.256637061 \ldots \times 10^{-6}\ mkg/C^2$ magnetic constant
$c_1 = 3.741771852 \ldots \times 10^{-16}\ m^4kg/s^3$ 1st radiation constant
$\sigma_e = 6.6524616(18) \times 10^{-29}\ m^2$ electron Thomson cross section
$G_0 = 7.748091729 \ldots \times 10^{-5}\ sC^2/m^2kg$ conductance quantum
$a_0 = 5.2917721092(17) \times 10^{-11}\ m$ Bohr electron radius
$m_u = 1.66053906660(50) \times 10^{-27}\ kg$ atomic mass constant
$Z_0 = 3.76730313668(57) \times 10^{2}\ m^2kg/sC^2$ characteristic impedance
$\sigma = 5.670374419 \times 10^{-8}\ kg/s^3K^4$ Stefan-Boltzmann constant
$N_A = 6.02214076 \times 10^{23}\ 1/mol$ Avogadro constant
$R_K = 2.58128074434(84) \times 10^{4}\ m^2kg/sC^2$ von Klitzing constant
$E_h = 4.3597447222071(85) \times 10^{-18}\ m^2kg/s^2$ Hartree energy
$c = 2.99792458 \times 10^{8}\ m/s$ speed of light

$c_{1L} = 1.191042869(53) \times 10^{-16}\ m^4kg/s^3$ spectral radiance
$r_e = 2.8179403227(19) \times 10^{-15}\ m$ classical electron radius
$\mu_N = 5.050783699(31) \times 10^{-27}\ m^2C/s$ Nuclear magneton
$c_2 = 1.438776877 \ldots \times 10^{-2}\ m\ K$ 2nd radiation constant
$g_\mu = -2.00233184122(82)$ muon g-factor
$g_e = -2.00231930436256(35)$ electron g-factor
$R = 8.314462618\ m^2kg/s^2K\ mol$ molar gas constant
$\Phi_0 = 2.067833848 \ldots \times 10^{-15}\ m^2kg/sC$ magnetic flux constant
$q_c = 3.6369475516(11) \times 10^{-4}\ m^2/s$ quantum of circulation
$g_+ = +5.5856946893(16)$ proton g-factor
$R_\infty = 1.0973731568539(55) \times 10^7\ 1/m$ Rydberg constant
$\gamma_+ = 2.6752218744(11) \times 10^8\ s/kg\ C$ proton gyromagnetic ratio
$N_\mu = -9.6623647(23) \times 10^{-27}\ m^2C/s$ neutron magnetic moment
$F = 9.648533212 \ldots \times 10^4\ C/mol$ Faraday constant
$g_N = -3.82608545(90)$ neutron g-factor
$\alpha_G = 1.7518(21) \times 10^{-45}$ gravitational coupling constant
$\omega_c = 7.763441 \times 10^{20}\ 1/s$ Compton angular frequency
$S_{mi} = 4.419 \times 10^9\ kg/sC$ Schwinger magnetic induction
$G = 6.67384(80) \times 10^{-11}\ m^3/s^2kg$ gravitational constant
$k_B = 1.380649 \times 10^{-23}\ m^2kg/s^2K$ Boltzmann constant
$r_+ = 8.751(61) \times 10^{-16}\ m$ proton radius

Where the digits in the parentheses define the measurement error in the preceding two digits (e.g. $8.751(61)$ means 8.751 ± 0.061), and the neutrino masses are only known to be non-zero.

This book tells the story of these numbers, revealing that all the actions of quantum field theory and general relativity are mapped by the combinatorial logic of the minimal self-balanced manifold—the hyperbolic figure eight knot. That is, in this book we notice for the first time that the ideal minimum geometry, the one that defines the minimal possible self-persistent stage, is responsible for setting the stage of physical reality—and for giving it all of its constructive parameters.

The layout of this book is as follows.

In Chapters 0-4 we fully decompose the hyperbolic figure eight knot under ideal partition balance.

In Chapter 5 we discover that the boundary conditions of this decomposition precisely predict the Planck constants, the mysterious limits (boundaries) of the 5 fundamental *dimensions of measure* in physics (time, space, charge, mass, and temperature).

In Chapters 6-9 we observe that the external facing boundaries of that balance connect under hyperbolic vortex arrangement, and that this connection precisely partitions the Planck charge and Planck mass boundaries into the charge and mass assignments that define the fundamental particles of matter—predicting all of their values geometrically.

In Chapter 10 we find that the unique Planck intersections balanced by this minimally partitioned geometry define the constants of Nature.

In Chapters 11-12 we notice that the Planck union is dually hyperbolic. That is, the Planck boundaries are maintained under ideal hyperbolic connection (defined by e, the base of natural logarithms) while terminally ideally hyperbolically partitioning (defined by the gamma function).

And in Chapters 13-15 we discover that the Riemann zeta function, the octonions, and the logic of calculus are unique combinatorial features of this minimum partition balance.

In summary, the unique properties of the minimal self-balanced geometry elegantly define the constructive parameters of reality. As in the "impossible to predict" parameters at the heart of physics, the unexplained missing soul of our best current story of reality's structure. The balance created by the division boundary of hyperbolic figure eight knot defines the minimal *arena of persistence*. The geometry of that simplest self-balance defines the shape of reality. And the division parameters of that shape define the fundamental properties of physics.

Chapter 0: the simplest manifold

In the mid 1970's Robert Riley and Troels Jørgensen independently discovered that the figure eight knot emits a complement with hyperbolic structure. This was the first known example of a hyperbolic knot—a closed boundary formed under hyperbolic balance.

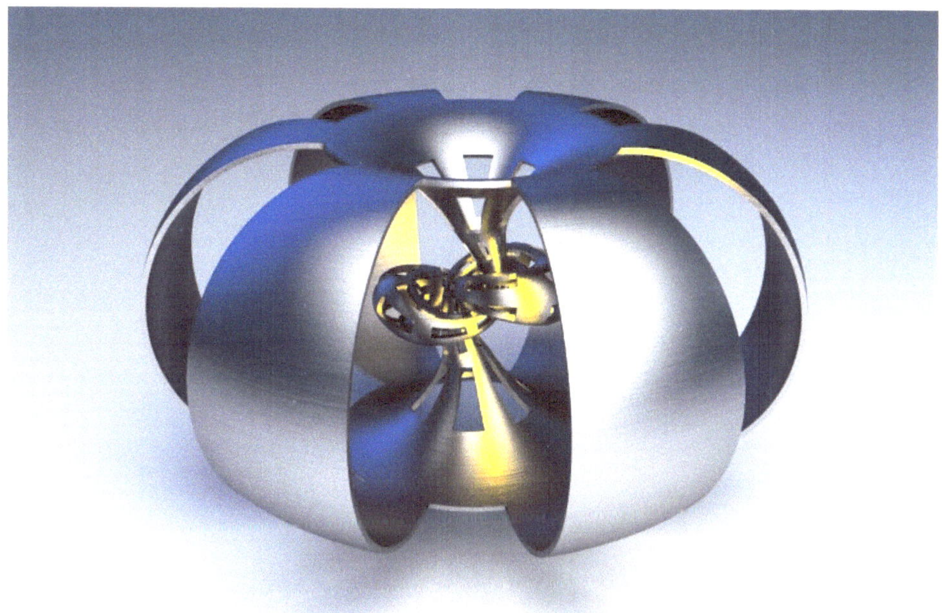

The complement of the hyperbolic figure-eight knot (portrayed here with cut-outs to allow visibility) is the finite volume bounded by the simple closed geodesics that trace out the figure-eight knot.

The hyperbolic figure eight knot is a double cover of the Gieseking's manifold, the simplest among all non-compact hyperbolic 3-manifolds.[1] Its internal volume complement (equal to twice the Gieseking's constant (G_{Gi})) defines the smallest possible volume complement (V_{fe}).

[1] A manifold is a geometry that self-closes, one whose boundary has been removed. For example, to transform a rectangle (a 2-dimesnional geometric form) into a manifold we fold it around on itself one way to form a cylinder, removing 2 of the boundaries by connecting them, then we twist that tube into a closed torus by connecting the ends of the cylinder, removing the remaining boundaries.

$$V_{fe} = 2G_{Gi} = 2.02988321281930\ldots$$

This minimally persistent balance separates a finite domain into 2 finite regions of momentum, a volume whirling about inside the complement division boundary and a finite volume outside that boundary moving under mirrored action.

Together, the internal and external actions of this balance define the minimal persistent geometry (universe). This minimal self-balanced stage defines the simplest possible perpetual balance of boundary conditions. That is to say, *persistence* is *minimally* obtained when a filament (a finite wave element) becomes balanced under local cyclic action by dividing over the division boundary of the hyperbolic figure eight knot, the minimal persistent geometry.

When William Thurston looked at this ideal geometry, he discovered that its internal complement decomposes into a union of 2 regular ideal hyperbolic tetrahedra. He then famously formulated his geometrization conjecture, claiming that all 3-manifolds admit a certain kind of geometric decomposition involving 8 geometries, most of which are hyperbolic. Grigori Perelman proved this conjecture in 2002-2003.

Thurston's decomposition was a brilliant addition to our knowledge of the hyperbolic figure eight knot, but it didn't tell us everything there is to know. That is, it wasn't explicit enough to define every feature of this minimum geometry.

For a complete understanding of the hyperbolic figure eight knot's geometric properties, to sharply define its internal and external projections, its boundary conditions and the balance of intersections maintained by those boundaries, we need to explicitly decompose its full partition balance (internally and externally).

Let's begin by generally decomposing this geometry into its 5 unique balances.

Note: The internal and external geometries of this balanced system are maintained under Möbius connection. That is, the same circles connect into 2 different (internal and external) constructive geometries. In the figures that follow, to make the internal/external partition balance easier to compare, I have connected the circles in both ways (into their internal and external geometric expressions) and placed them side by side. The geometry on the left shows what the division boundary looks like to an *internal* observer—inside the minimum volume complement (the hyperbolic figure eight knot complement boundary), while the geometry on the right shows what the division boundary looks like from the point of view of an observer within the *external* domain (the n-hypersphere of maximal volume).

Chapter 1: the minimum partition balance

The action of the hyperbolic figure eight knot is 5 dimensional—it decomposes into 5 unique geometrically balanced actions (balance 0-4).

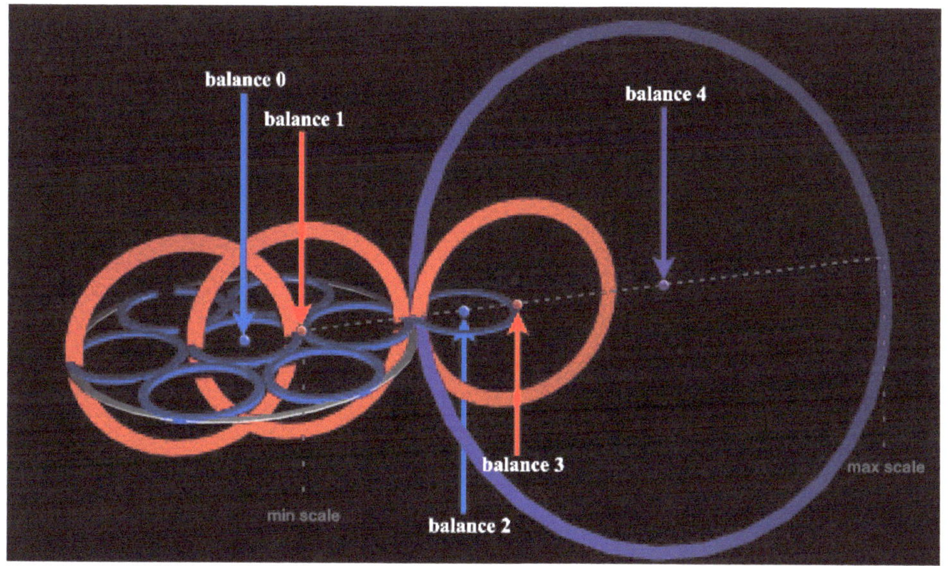

The partition balance of the hyperbolic figure eight knot balances 2 internal actions (balance 0-1) against 3 unique external actions (balance 2-4).

general decomposition map of the hyperbolic figure eight knot

$(\text{inner action})_4{}^{-1}(\text{outer rotation})_4{}^{-1} = d_4{}^{-1}$	$(\text{boundary } 4)^{-1}$
$(\text{the external reference}) = 1$	balance 4
$(\text{inner action})_3(\text{outer rotation})_3 = d_3$	balance 3
$(\text{inner action})_2(\text{outer rotation})_2 = d_2$	balance 2
$(\text{inner action})_4(\text{outer rotation})_4 = d_4$	boundary 4
$(\text{inner action})_1(\text{outer rotation})_1 = d_1$	balance 1
$(\text{inner action})_0(\text{outer rotation})_0 = d_0$	balance 0

Where the union of balance 0 and balance 1 defines the internal double-cover domain, balance 4 (created by the union of balance 2 and balance 3) defines the external split-cover domain, and $d_k =$ the number of derangements involved in each balance.

The balance point of the external domain (balance 4) prescribes the unitary scale of this decomposition—the 1—the scale of stability projected under this 5-part action. Under this ideal inversion balances 2-4 are the mirrored (Möbius) actions of balances 0-1.

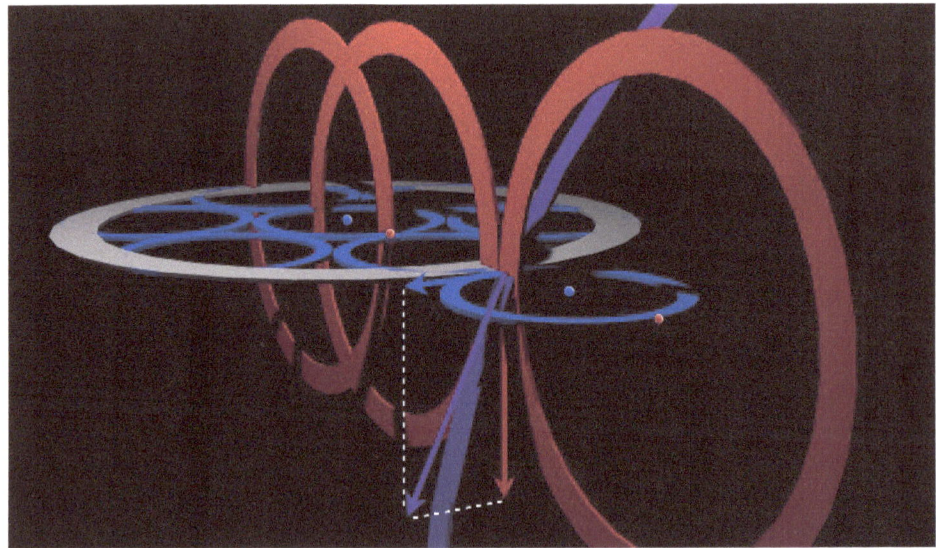

The combined action of balance 2 and 3 (blue and red arrows) rotates the external domain under balance 4 (purple).

Since this balance defines a finite localized action, the rotations of its internal and external parts must perfectly close. Under ideal *internal* split factorization $\sqrt{!n - bn}$ the *external* partitions close as *doubled* and *squared* arrangements of those internal factorizations $2\left(\sqrt{!n - bn} + k\right)^2$. Since this geometry involves 3 unique external actions, that external arrangement comes in 3 phases k, where $k = (-1,0,1)$.

external

$$d_3 = 2\left(\sqrt{!n - bn} - 1\right)^2 = 2(3-1)^2 = 8$$

$$d_2 = 2\left(\sqrt{!n - bn} \pm 0\right)^2 = 2(3 \pm 0)^2 = 18$$

$$d_4 = 2\left(\sqrt{!n - bn} + 1\right)^2 = 2(3+1)^2 = 32 = 2^n$$

internal

$$d_1 = bn = 35$$

$$d_0 = !n = 44$$

Where $n = 5$ the number of unique rotations partitioning the balance of the hyperbolic figure eight knot, $!n = 44$ the number of derangements available to 5 rotations, $b = 7$ the break in scale symmetry between the first and second balanced scales maintained under this action (see Chapter 3), and the 3 external phases $(-1,0,1)$ define the 3 surfaces of hyperbolic geometry encoded by the quadratic form:

$$x^2 + y^2 - z^2 = k$$

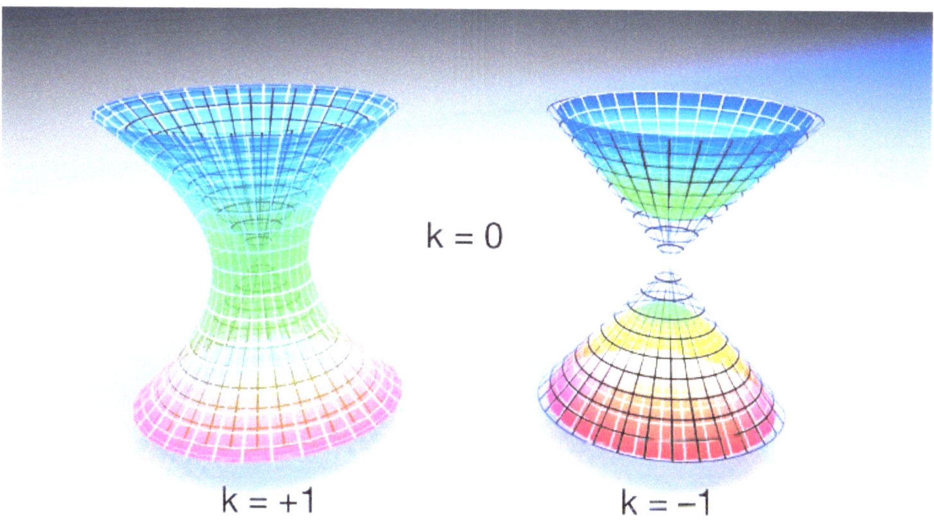

The 3 external surfaces of hyperbolic geometry: $k = +1$ defines the one-sheeted hyperploid, $k = 0$ defines the *null cone,* the conic division boundary of the projective plane, and $k = -1$ defines the two-sheeted hyperploid.

This minimum symmetry constraint (being maintained under finite closed rotation via split-symmetric division, under closed external quadratic connection) allows us to completely specify the right-hand side of the hyperbolic figure eight knot's decomposition map.

To sharpen this map further, we note that no matter how complex the interior action of a particular balance is, its exterior expression will always be just a smooth rotation of some magnitude. In other words, every $(\text{outer rotation})_k = e^{\Phi_k}$.

decomposition map of the hyperbolic figure eight knot

$(\text{inner action})_4{}^{-1} e^{-\phi_4} = 2^{-n}$	$(\text{boundary } 4)^{-1}$
$(\text{external reference}) = 1$	balance 4
$(\text{inner action})_3\, e^{\phi_3} = 2(3-1)^2$	balance 3
$(\text{inner action})_2\, e^{\phi_2} = 2(3 \pm 0)^2$	balance 2
$(\text{inner action})_4\, e^{\phi_4} = 2(3+1)^2 = 2^n$	boundary 4
$(\text{inner action})_1\, e^{\phi_1} = bn$	balance 1
$(\text{inner action})_0\, e^{\phi_0} = !n$	balance 0

Where ϕ_0, ϕ_1, ϕ_2, ϕ_3, & ϕ_4 are the magnitudes of rotation keeping each external balance $(\text{outer rotation})_k = e^{\phi_k}$, $k = \{0,1,2,3,4\}$, the union of balance 0 and balance 1 defines the internal double-cover domain (the internal complement), balance 4 (created by the union of balance 2 and balance 3) defines the external split-cover domain (the n-hypersphere of maximal volume), $n = 5$ the number of unique rotations partitioning the balance of the hyperbolic figure eight knot, $!n = 44$ the number of derangements available to 5 rotations, and $b = 7$ the break in scale symmetry between the first and second balances (see Chapter 3).

To complete this decomposition map we must specify its balance of inner actions. Let's begin by defining the absolute minimum cyclic expression (balance 0).

Chapter 2: the minimum limit of persistence

Let the minimum balance in the hyperbolic figure eight knot (balance 0) define the complete *derangement* ($!\,n$) of its 5 unique rotations ($n = 5$).[2] And let the geometry of this minimum cyclic action be defined as a circularly closed action πr^2 trivially balanced against an external counter-rotation e^{ϕ_0}.

$$\pi r^2 e^{\phi_0} = !\,n \qquad \text{balance 0}$$

Where $n = 5$ the number of unique rotations partitioning this balance, $!\,n = 44$ the number of derangements available to those 5 rotations, and $\phi_0 =$ the magnitude of the external rotation balancing this minimum cyclic action.

To represent this minimal circle being maintained under hyperbolic construction (factoring into rotations that counterbalance under ideal complex division—splitting into equal but opposite internal rotations, each absorbing half the input) we set r equal to the hyperbolic sine function.

$$\frac{1}{2}(e^x - e^{-x}) = sinh(x) = r$$

And we set the argument of this function (x) equal to the minimal representation of square split division—a number that constructively possesses complex four-fold symmetry.

$$x = \left(\frac{1}{2}\right)^2 = \left(-\frac{1}{2}\right)^2 = -\left(\frac{1}{2}i\right)^2 = -\left(-\frac{1}{2}i\right)^2$$

$$\frac{1}{2}\left(e^{\left(\frac{1}{2}\right)^2} - e^{-\left(\frac{1}{2}\right)^2}\right) = sinh\left(\left(\frac{1}{2}\right)^2\right) = r$$

Plugging this r into our equation, we arrive at a precise characterization of the minimum cyclic limit of *persistence*—the smallest cyclic action maintained by the square split-division balance of the hyperbolic figure eight knot.

---

[2] A derangement is a combinatorial permutation that has no fixed points. That is, it is a balanced rearrangement whose parts all play active roles.

$$\pi\left(\sinh\left(\left(\frac{1}{2}\right)^2\right)\right)^2 e^{\phi_0} = !\boldsymbol{n} \qquad \text{balance } 0$$

Where $\phi_0 = 5.39125836832313\ldots$ the magnitude of the external rotation maintaining balance 0, $sinh(x) =$ the hyperbolic sine function, $\boldsymbol{n} = 5$ the number of unique rotations partitioning this balance, and $d_0 = !\boldsymbol{n} = 44$ the number of derangements available to 5 rotations.

Since this equation defines the minimum cyclic action within the minimum self-balanced geometry, it holds the honor of defining the *ultimate* boundary condition; the absolute minimum limit of measure, beyond which the possibility for *persistence* itself is operationally cut off. In other words, this equation defines the minimum boundary of time.

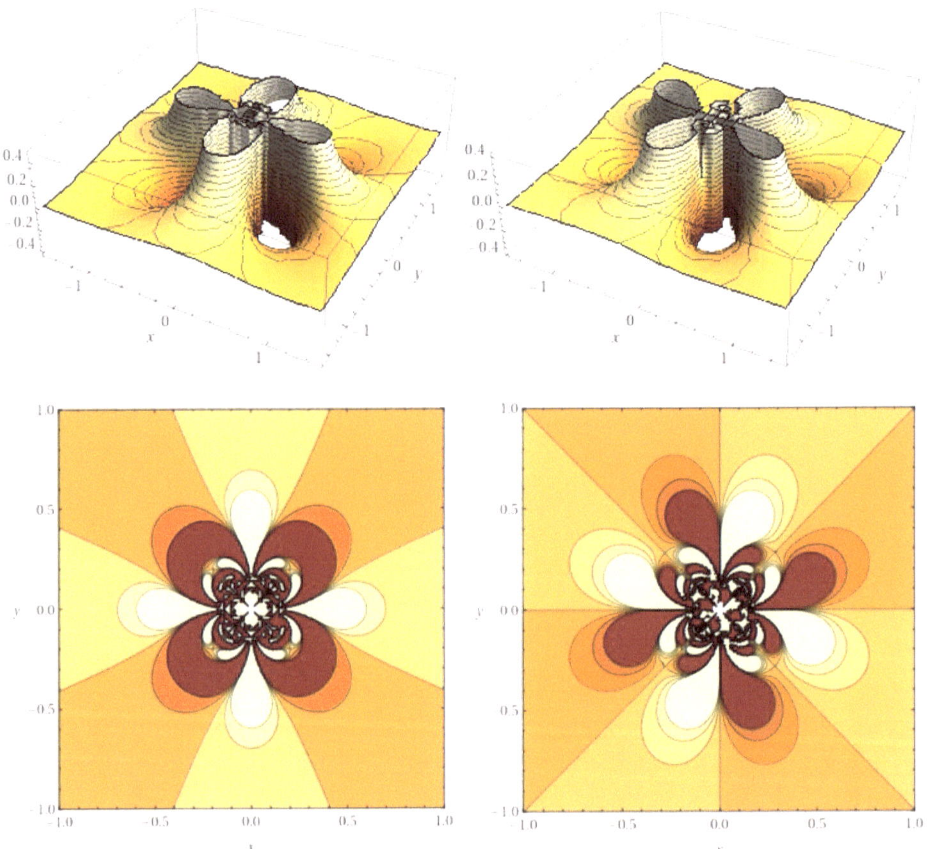

Graph 1: real (left) and imaginary (right) plots of the internal action of balance 0 under inverse-complex argument.

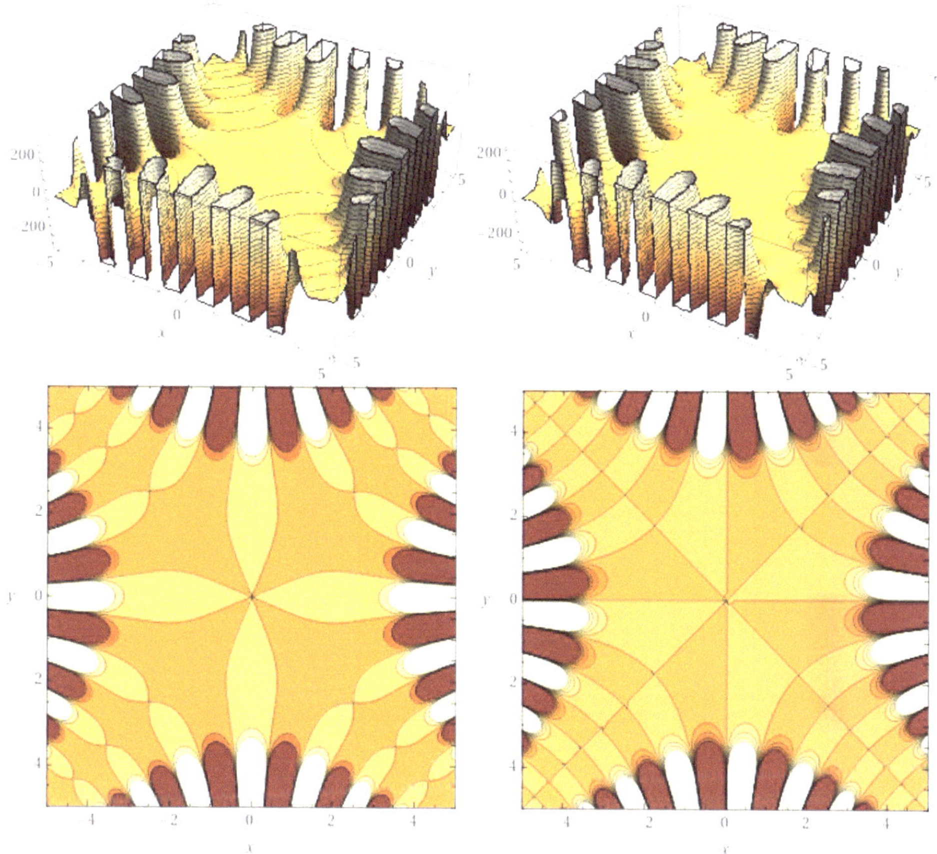

Graph 2: real (left) and imaginary (right) plots of the internal action of balance 0 under complex argument.

Equivalent expressions for the minimum limit of persistence include:

$$\pi\left(i\,sin\left(\left(\frac{\sqrt{i}}{2}\right)^2\right)\right)^2 e^{\Phi_0} = !\,\boldsymbol{n} \qquad \text{balance 0}$$

$$\frac{\pi}{2}\left(\frac{1}{2}\left(e^{-1/2}+e^{1/2}\right)-1\right)e^{\Phi_0} = !\,\boldsymbol{n} \qquad \text{balance 0}$$

and

$$cosh\left(\left(\frac{1}{2}\right)^2\right)\left|\Gamma\left(\frac{1}{2}+\frac{i}{4\pi}\right)\right|^2\left(sinh\left(\left(\frac{1}{2}\right)^2\right)\right)^2 e^{\Phi_0} = !\,\boldsymbol{n} \qquad \text{balance 0}$$

Chapter 3: the first consequence

Under circular construction, the next sized circle that can be constructed is always composed of 7 circles (6 outer circles surrounding 1 inner circle) and the negative space that divides them. Therefore, the break in scale-symmetry between the first and second scales of any circular construction is $\boldsymbol{b} = 7$.

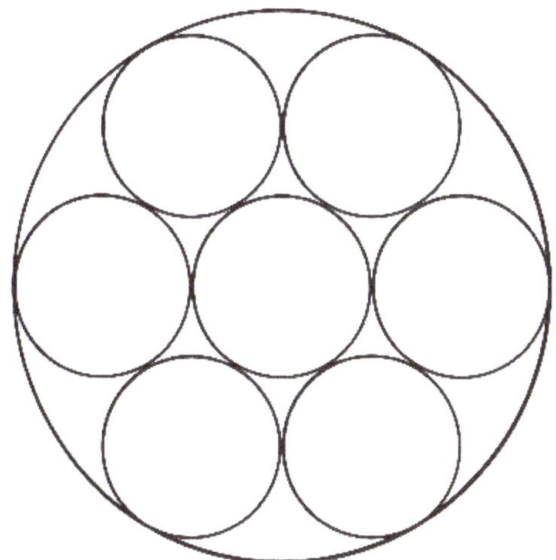

$$\frac{\text{\# of circles included on } 2^{\text{nd}} \text{ scale}}{\text{\# of circles included on } 1^{\text{st}} \text{ scale}} = \boldsymbol{b} = 7$$

With this initial constructive arrangement in mind, let's zoom out to the next balance of the hyperbolic figure eight knot (balance 1), whose inner action is the operationally rotated inverse action of balance 0, adjusted for the break in scale symmetry between the 2 balances. This balance maintains a new external rotation (ϕ_1) that rotates (cycles through) $\boldsymbol{bn} = 35$ of the fundamental $!\boldsymbol{n} = 44$ derangements.

$$\left(sinh\left(sinh\left(\frac{1}{\boldsymbol{b}}\right)\right)\right)^{-1} e^{\phi_1} = \boldsymbol{bn} \qquad \text{balance 1}$$

Where $\phi_1 = 1.61625918175645\ldots$ is the magnitude of the external rotation maintaining balance 1, $sinh(x) =$ the hyperbolic sine function, $\boldsymbol{b} =$ the break in scale symmetry between the first and second scales of balanced action in the hyperbolic figure eight knot, and $d_1 = \boldsymbol{bn} = 35$ the number of derangements participating in balance 1.

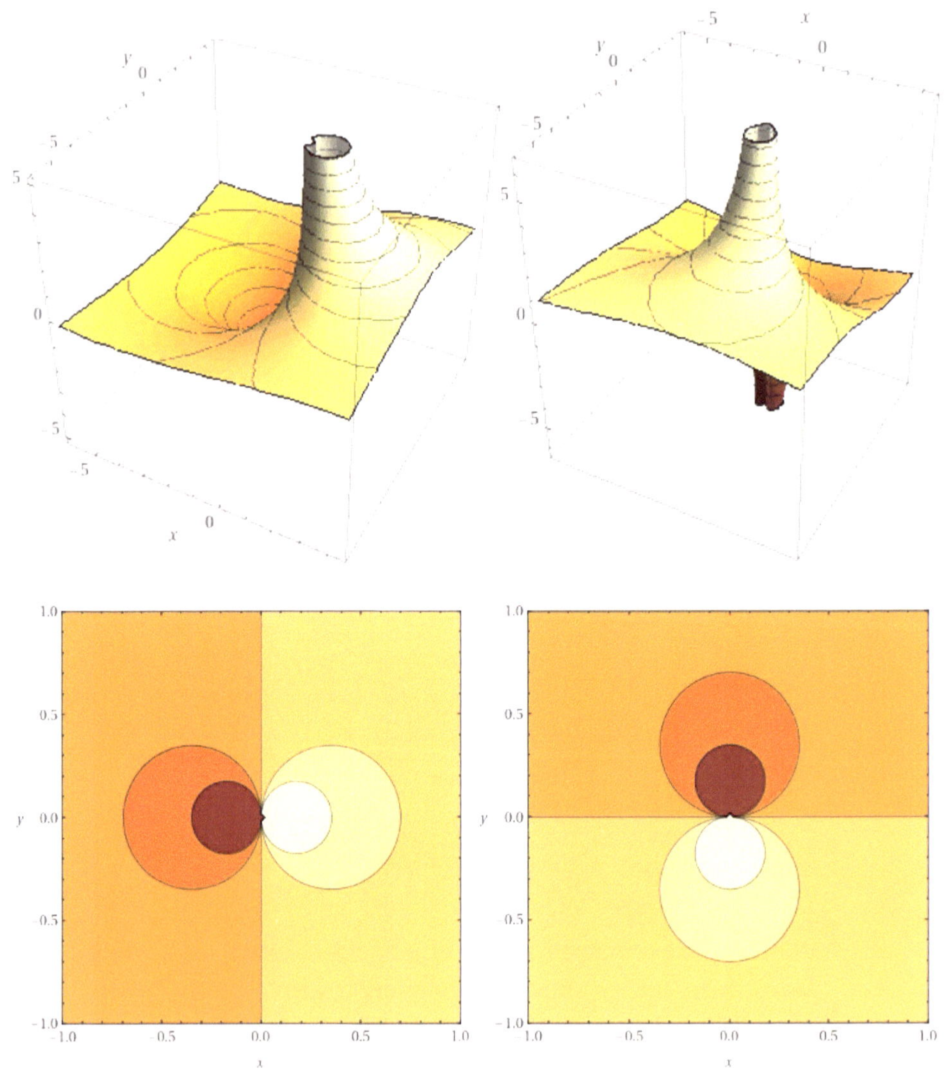

Graph 3: real (left) and imaginary (right) plots of the internal action of balance 1 under complex argument.

Balance 1 can be equivalently written:

$$\left(\frac{1}{2}\left(e^{\left(\frac{1}{2}\left(e^{1/b}-e^{-1/b}\right)\right)}-e^{\left(\frac{1}{2}\left(e^{-1/b}-e^{1/b}\right)\right)}\right)\right)^{-1}e^{\phi_1}=\boldsymbol{bn}$$

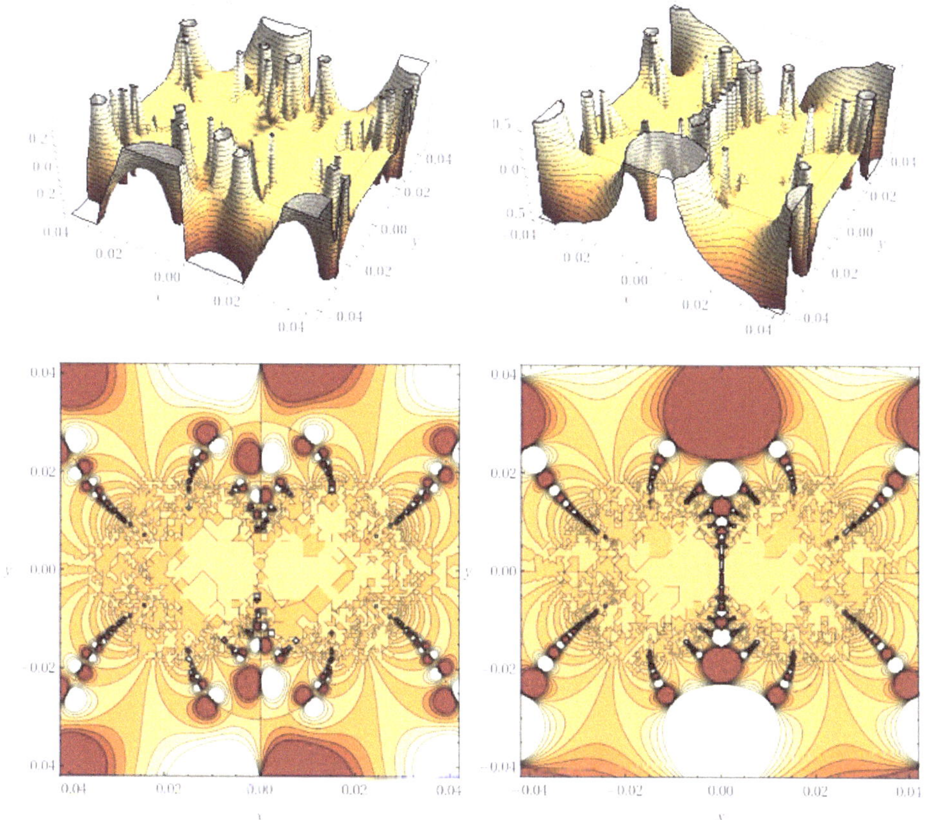

Graph 4: real (left) and imaginary (right) plots of the internal action of balance 1 under inverse-complex argument, zoomed in.

The n unique partitions of this geometric balance trivially connect under square arrangement (n^2) while orthogonally factoring under ideal hyperbolic balance ($\Gamma(n)$).

$$b\, cosh(\log b) = n^2$$

$$b\, sinh(\log b) = \Gamma(n)$$

Where $cosh(x) =$ the hyperbolic cosine function, $sinh(x) =$ the hyperbolic sine function, $\log(x) =$ the natural logarithm function, $\Gamma(x) =$ the gamma function which encodes hyperbolically balanced partitions, $n = 5$ the number of unique rotations partitioning the hyperbolic figure eight knot and, $b = 7$ the break in scale symmetry of this balance.

Chapter 4: the external balances

The external domain of the hyperbolic figure eight knot connects under ideal hyperbolic vortex arrangement (see Chapter 6).

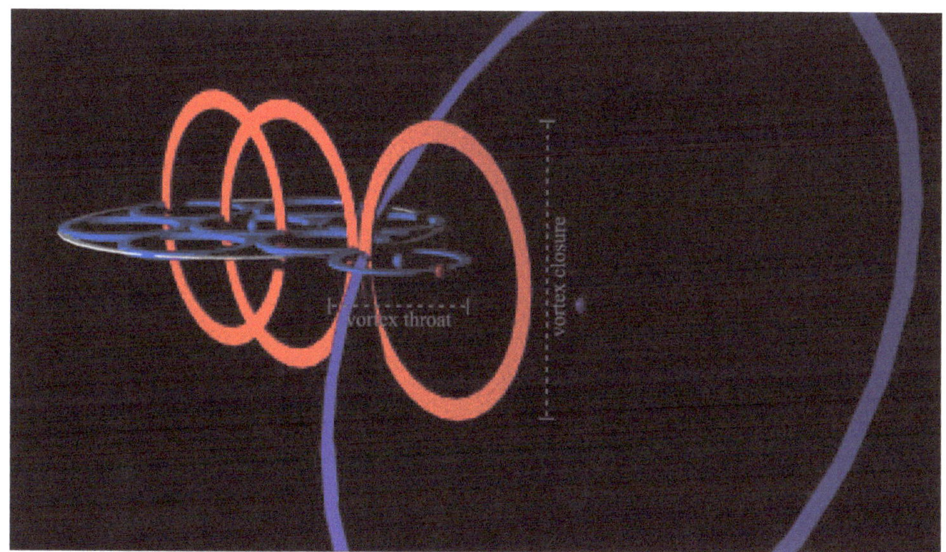

The boundaries of the 2 external balances (balance 2 and 3) form the vortex's throat and its terminal limit—the blue and red circles within the external (purple) domain.

Balance 2 forms the vortex's throat, defining where the n unique partitions doubly-periodically divide (W_{We}), while factoring along the break in scale symmetry of the geometry $\sqrt{b}$ and wrapping axisymmetrically around a single pole $(3)^{1/3}$.

$$\frac{n}{W_{We}\sqrt{b}\,(3)^{1/3}}e^{\phi_2} = 2(3 \pm 0)^2 \qquad \text{balance 2}$$

Where $\phi_2 = 1.87554596713962\ldots$ is the magnitude of the external rotation maintaining balance 2, $n = 5$ the number of unique rotations partitioning the balance of the hyperbolic figure eight knot, $d_2 = 2(3)^2 = 18$ the number of derangements participating in balance 2, and $W_{We} = \frac{1}{2}\sigma(1|1,i)$ is the Weierstrass constant—the unitary balance of the Weierstrass sigma function.

$$W_{We} = (\,2^{2n}\, e^{\pi}\,)^{1/8} \frac{\Gamma\left(\frac{1}{2}\right)}{\left(\Gamma\left(\left(\frac{1}{2}\right)^{2}\right)\right)^{2}}$$

Where $\Gamma(x) =$ the gamma function, and $\Gamma\left(\frac{1}{2}\right) = \sqrt{\pi}$.

The Weierstrass constant defines the harmonic double cover projection of the n-dimensional hypersphere, internally partitioning into 8 divisions maintained under lemniscate balance (divided by $2\sqrt{2}\,L$).

$$\mathrm{W_{We}} = (\,2^{2n}\, e^{\pi}\,)^{1/8} \frac{1}{2\sqrt{2}\,L}$$

Where $L =$ the lemniscate constant, s is the arc length of a unitary lemniscate (with $a = 1$),

$$L = \frac{1}{2}s \qquad\qquad s = \frac{1}{\sqrt{2\pi}}\left(\Gamma\left(\left(\frac{1}{2}\right)^{2}\right)\right)^{2}$$

and $e^{\pi} =$ the volume sum of all even-dimensional hyperspheres, which follows from the fact that the equations for the volume and surface areas of n-dimensional hyperspheres of radius r are:

$$V_n(r) = \frac{\pi^{n/2}}{\Gamma\left(\frac{n}{2}+1\right)}\, r^{n} \qquad\qquad S_{n-1}(r) = \frac{2\pi^{n/2}}{\Gamma\left(\frac{n}{2}\right)}\, r^{n-1}$$

$$\lim_{n\to\infty} \frac{\pi^0}{\Gamma(1)} + \frac{\pi^1}{\Gamma(2)} + \frac{\pi^2}{\Gamma(3)} + \cdots + \frac{\pi^n}{\Gamma(n+1)} = e^{\pi}$$

$$\lim_{n\to\infty} \frac{\pi^0}{0!} + \frac{\pi^1}{1!} + \frac{\pi^2}{2!} + \frac{\pi^3}{3!} + \cdots + \frac{\pi^n}{n!} = e^{\pi}$$

Where $\Gamma(x) =$ the gamma function and $n!$ is the factorial function.

This ideal partition balance has a few orthogonal components. That is, alongside the unitary Weierstrass sigma function comes its associated

Weierstrass elliptic function with equianharmonic, lemniscatic, and pseudo-lemniscatic half-periods.

equianharmonic half-periods

$$\omega_1 = \frac{\Gamma^3\left(\frac{1}{3}\right)}{4\pi}\left(\frac{1}{2}\left(i\sqrt{3}+1\right)\right) \qquad \omega_2 = \frac{\Gamma^3\left(\frac{1}{3}\right)}{4\pi}$$

lemniscatic half-periods

$$\frac{L}{2\sqrt{2}} \qquad \frac{L}{2\sqrt{2}}\,i$$

pseudo-lemniscatic half-periods

$$\frac{L}{4}(\,i+1) \qquad \frac{L}{4}(\,i-1)$$

The lemniscate plays a central (quadropoly orthogonal) role in this external balance because "The lemniscate is the inverse curve of the hyperbola with respect to its center." [3]

The lemniscate can also be generated as the envelope of circles centered on a rectangular hyperbola and passing through its center.

---

[3] Wells, D. *The Penguin Dictionary of Curious and Interesting Geometry*. London: Penguin, pp. 139-140, 1991. Figure and caption from mathworld.wolfram.com/Lemniscate.html

Instead of writing balance 2 in terms of the Weierstrass constant, or the lemniscate constant, let's write it in terms of the gamma function, to highlight how it is maintained under ideal square/split hyperbolic balance.

$$\frac{n}{\sqrt{b\pi}\ (3)^{1/3}}\left(2^{2n}\, e^{\pi}\right)^{-1/8}\left(\Gamma\left(\left(\frac{1}{2}\right)^{2}\right)\right)^{2} e^{\phi_2} = 2(3 \pm 0)^2 \qquad \text{balance 2}$$

Where $\phi_2 = 1.87554596713962\ldots$ is the magnitude of the external rotation of balance 2, $\Gamma(x) =$ the gamma function, which encodes hyperbolically balanced partitions, and $d_2 = 2(3 \pm 0)^2 = 18$ the number of derangements participating in balance 2.

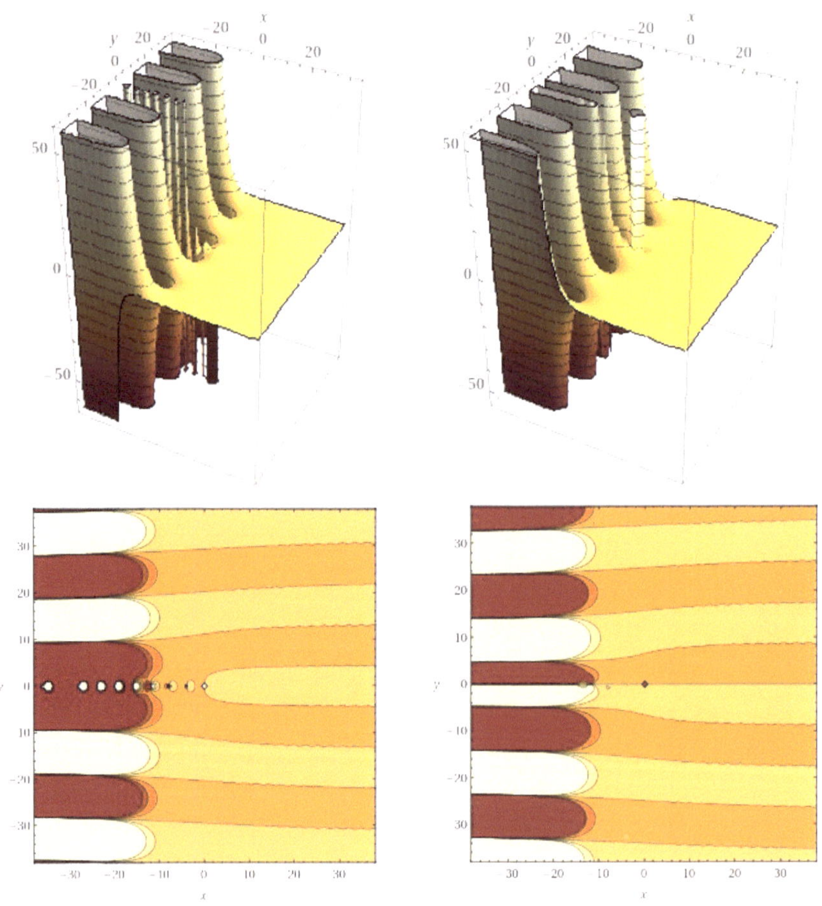

Graph 5: real (left) and imaginary (right) plots of the internal action of balance 2 under dual complex argument.

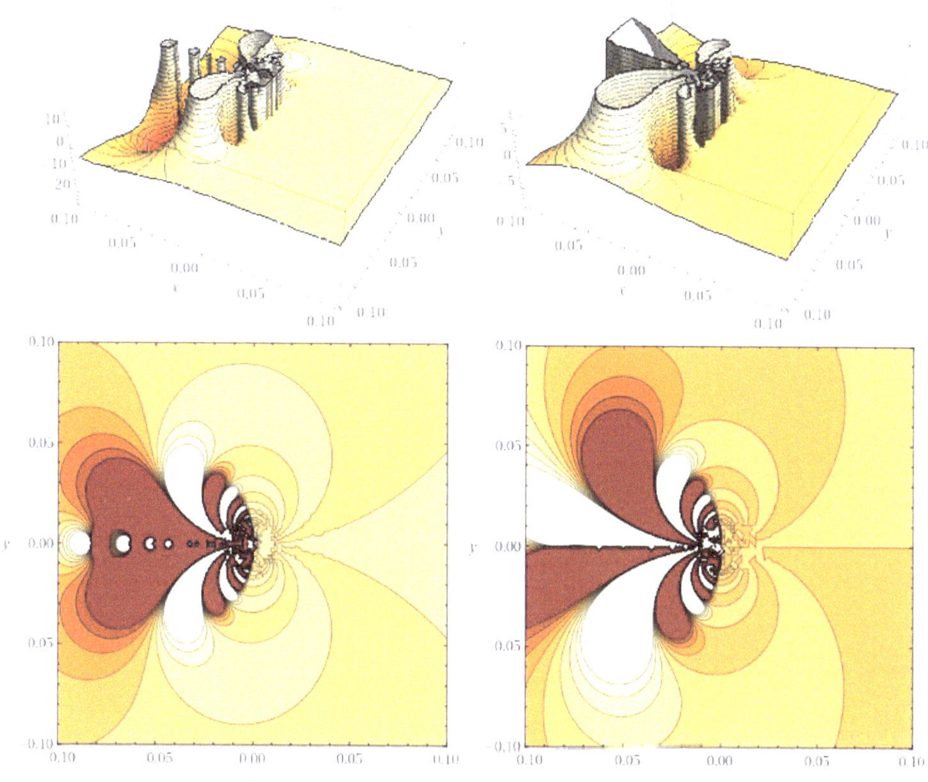

Graph 6: real (left) and imaginary (right) plots of the internal action of balance 2 under dual inverse-complex argument. "the partition butterfly"

Balance 3 defines the opposite end of hyperbolic vortex, where the external rotations of this division balance circularly close ($2\pi\, n$) while squarely balancing on the external *circular* split division of the hyperbolic figure eight knot.

$$2\pi\, n \left(cos\left(\frac{b}{n} \right) \right)^2 e^{\phi_3} = 2(3-1)^2 = 2^3 \qquad \text{balance 3}$$

Where $\phi_3 = 2.17642683817579\ldots$ is the magnitude of the external rotation maintaining balance 3, $cos\left(\frac{b}{n}\right) =$ the external circular split division of the hyperbolic figure eight knot, and $d_3 = 2^3 = 8 =$ the number of derangements participating in balance 3.

$$cos\left(\frac{b}{n} \right) = \frac{1}{2}\left(e^{-\left(\frac{b}{n}\right)i} + e^{\left(\frac{b}{n}\right)i} \right)$$

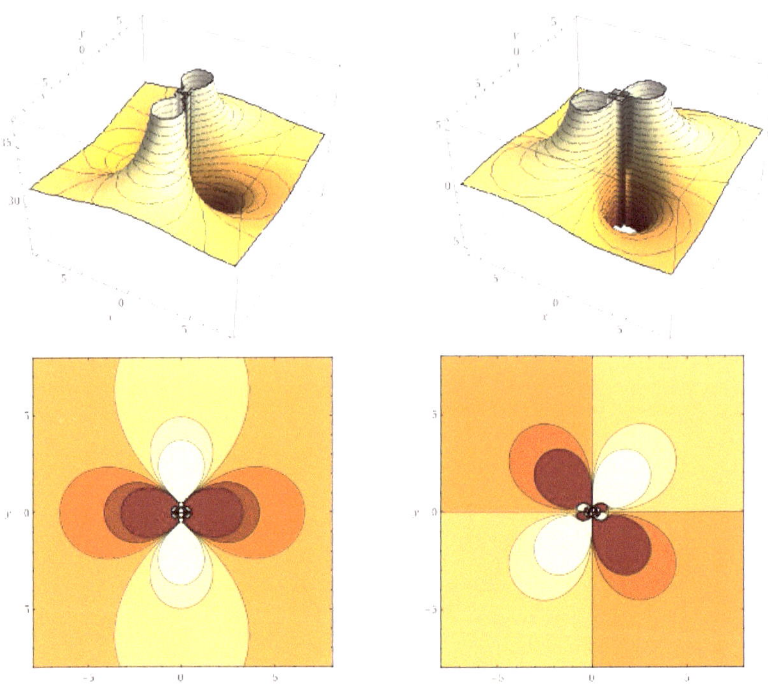

Graph 7: real (left) and imaginary (right) plots of the internal action of balance 3 under inverse-complex argument. "split figure eights"

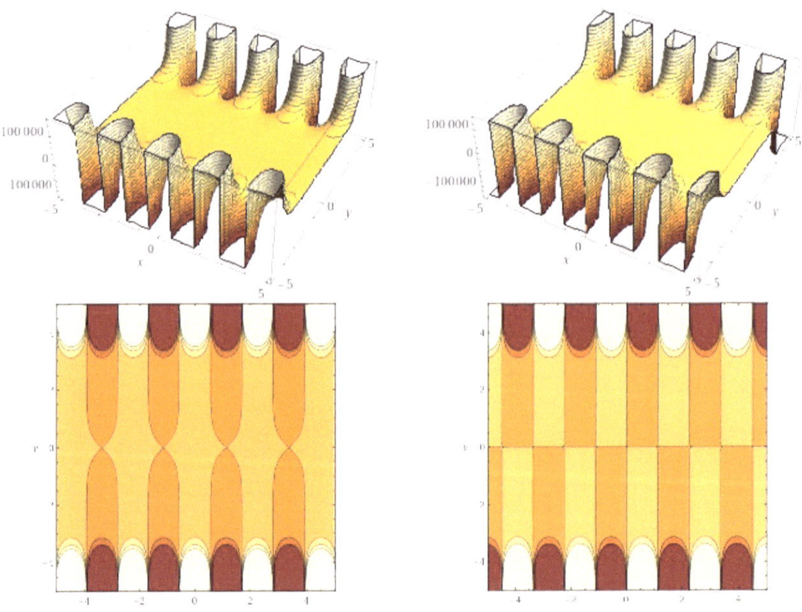

Graph 8: real (left) and imaginary (right) plots of the internal action of balance 3 under complex argument.

Balance 4 defines the most external projection of hyperbolic figure eight knot. The balance point of this external domain (the central scale within boundary 4 (purple)) prescribes the unitary scale of this entire decomposition—the 1—the scale most stably projected under the 5-part action of the hyperbolic figure eight knot.

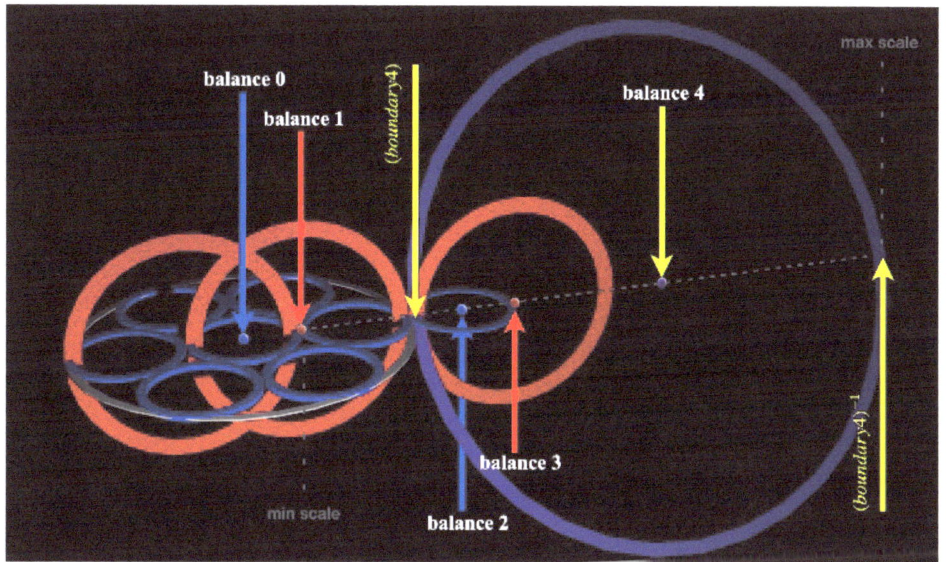

The internal and external boundaries of $balance\ 4$ *(*$boundary\ 4$ *&* $(boundary\ 4)^{-1}$*) are under inverse action.*

Boundary 4 maintains a union of inverse factors, $\left(\frac{1}{x}+x\right)=e^{\log x}+e^{-\log x}=2\,cosh(\log x)$, and squarely joins the internal hyperbolic split division of this balance $cosh\left(\frac{n}{2}\right)$ to its external circular split-division $cos\left(\frac{b}{n}\right)$.

$$2\,cosh(log\ \mathbf{b})\left(cosh\left(\frac{n}{2}\right)\right)^2\left(cos\left(\frac{b}{n}\right)\right)^2 e^{\phi_4}=2(3+1)^2=2^n \quad \text{boundary 4}$$

Where $\phi_4 = 1.41678698590795\ldots$ is the magnitude of the 4th external rotation, $cosh(x) =$ the hyperbolic cosine function, $\log(x) =$ the natural logarithm function, $cos(x) =$ the cosine function, $\boldsymbol{n} = 5$ the number of unique rotations partitioning the hyperbolic figure eight knot, $\boldsymbol{b} = 7$ the break in scale symmetry of the internal construction of this geometry (maintaining the dual inversion of this boundary), and $d_4 = 2^n = 32$ the number of derangements participating in the construction of boundary 4.

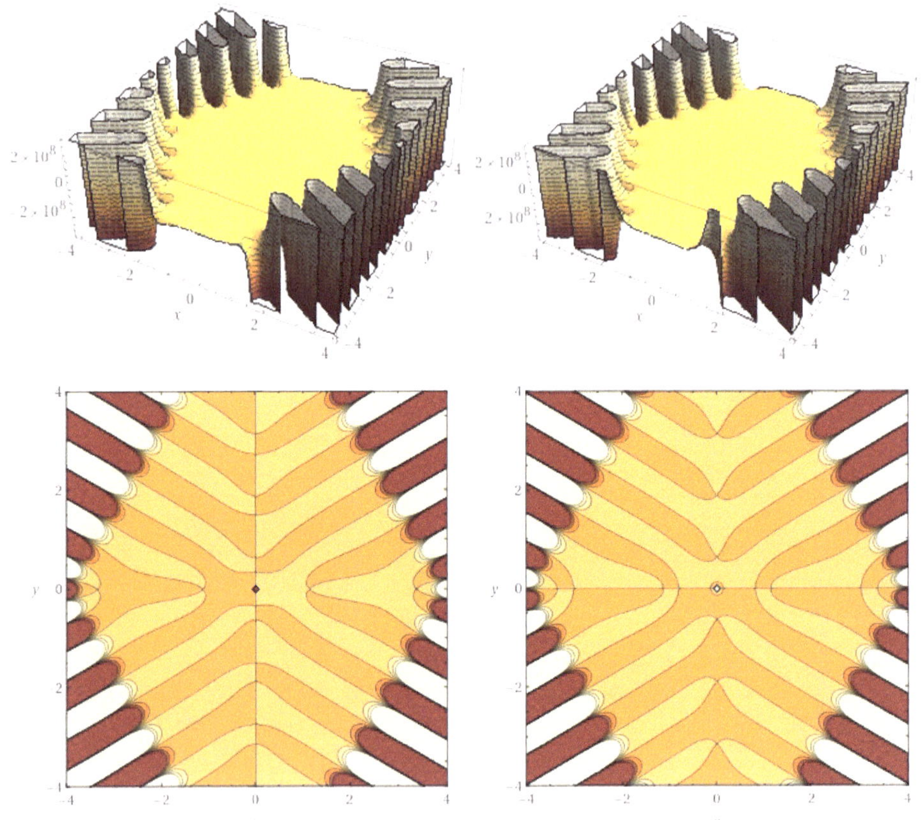

Graph 9: real (left) and imaginary (right) plots of the inner action of boundary 4 under triple complex argument.

$(\text{boundary } 4)^{-1}$ inversely expresses the action of boundry 4. This boundary defines the outer edge of action in this cyclically defined universe, the other boundary across which no action participating in this balance crosses.

$$(2\,cosh(log\,\boldsymbol{b}))^{-1}\left(cosh\left(\frac{\boldsymbol{n}}{2}\right)\right)^{-2}\left(cos\left(\frac{\boldsymbol{b}}{\boldsymbol{n}}\right)\right)^{-2}e^{-\phi_4} = 2^{-\boldsymbol{n}} \quad (\text{boundary } 4)^{-1}$$

Where $\phi_4 = 1.416786985907795\ldots$ is the magnitude of the 4th boundary's external rotation, $cosh(x)$ = the hyperbolic cosine function, $\log(x)$ = the natural logarithm function, $cos(x)$ = the cosine function, $\boldsymbol{n} = 5$ the number of unique rotations partitioning the balance of the hyperbolic figure eight knot, $\boldsymbol{b} = 7$ the break in scale symmetry of this geometry, and ${d_4}^{-1} = 2^{-\boldsymbol{n}}$ = the inverse derangements of this balance.

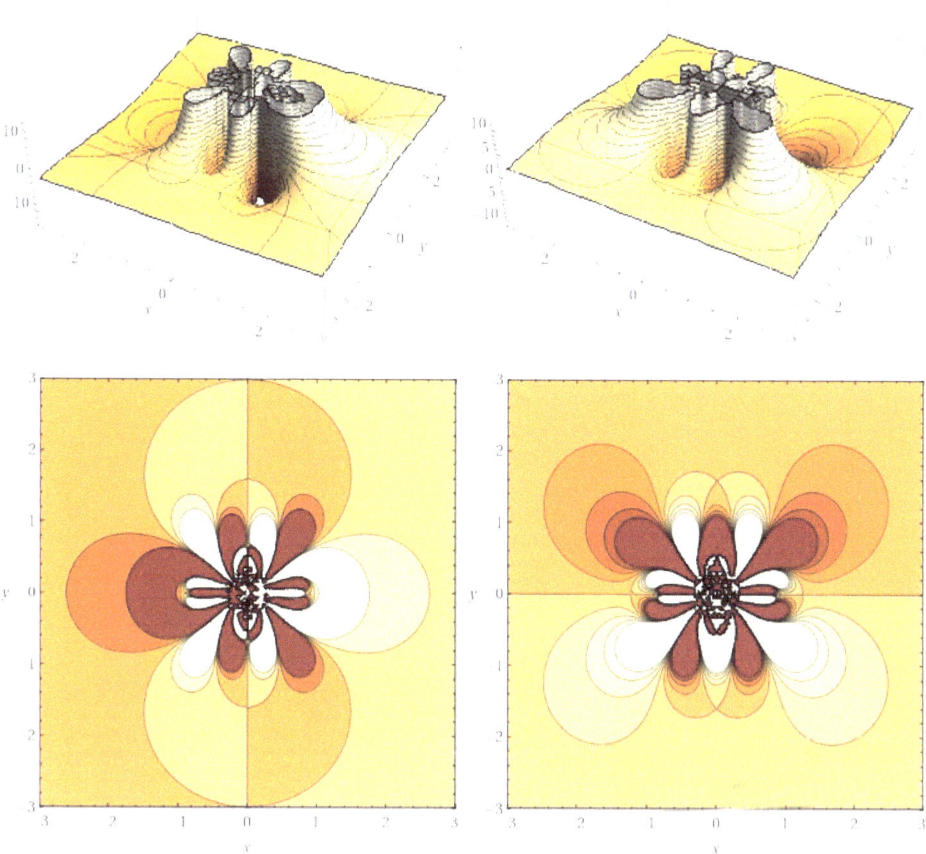

Graph 10: real (left) and imaginary (right) plots of the inner action of boundary 4 under triple inverse-complex argument.

By fully decomposing the minimum partition balance we have mapped the most elemental construction possible, defining the symmetrically closed balance of actions that compose the simplest possible persistent stage. To explore the intrinsic properties of that stage, let's look at its boundary conditions.

the complete partition balance of the hyperbolic figure eight knot

$$(2\,cosh(log\, b))^{-1}\left(cosh\left(\frac{n}{2}\right)\right)^{-2}\left(cos\left(\frac{b}{n}\right)\right)^{-2} e^{-\phi_4} = 2^{-n} \qquad (4^{th})^{-1}$$

$$(\text{external reference}) = 1 \qquad 4^{th}$$

$$2\pi\, n\left(cos\left(\frac{b}{n}\right)\right)^{2} e^{\phi_3} = 2(3-1)^2 = 2^3 \qquad 3^{rd}$$

$$\frac{n}{\sqrt{b\pi}\;(3)^{1/3}}\left(2^{2n}\, e^{\pi}\right)^{-1/8}\left(\Gamma\left(\left(\frac{1}{2}\right)^{2}\right)\right)^{2} e^{\phi_2} = 2(3\pm 0)^2 \qquad 2^{nd}$$

$$2\,cosh(log\, b)\left(cosh\left(\frac{n}{2}\right)\right)^{2}\left(cos\left(\frac{b}{n}\right)\right)^{2} e^{\phi_4} = 2(3+1)^2 = 2^n \qquad 4^{th}$$

$$\left(sinh\left(sinh\left(\frac{1}{b}\right)\right)\right)^{-1} e^{\phi_1} = b(n) \qquad 1^{st}$$

$$\pi\left(sinh\left(\left(\frac{1}{2}\right)^{2}\right)\right)^{2} e^{\phi_0} = !n \qquad 0^{th}$$

Where ϕ_k = the external rotation of the kth balance, $k = \{0,1,2,3,4\}$, d_k = the number of derangements participating in the kth balance, $n = 5$ the number of unique rotations partitioning the balance of the hyperbolic figure eight knot, $!n = 44$ the number of derangements available to those 5 rotations, $b = 7$ the break in scale symmetry internally maintained by this balance, $sinh(x)$ = the hyperbolic sine function, $cosh(x)$ = the hyperbolic cosine function, $cos(x)$ = the cosine function, $\Gamma(x)$ = the gamma function, which encodes hyperbolically balanced partitions, and $\log(x)$ = the natural logarithm function.

$\phi_4 = 1.4167869859079 \ldots$ $\qquad d_4 = 2(3+1)^2 = 2^n = 32$

$\phi_3 = 2.1764268381757 \ldots$ $\qquad d_3 = 2(3-1)^2 = 2^3 = 8$

$\phi_2 = 1.8755459671396 \ldots$ $\qquad d_2 = 2(3\pm 0)^2 = 18$

$\phi_1 = 1.6162591817564 \ldots$ $\qquad d_1 = bn = 35$

$\phi_0 = 5.3912583683231 \ldots$ $\qquad d_0 = !n = 44$

Chapter 5: the boundary conditions of physical reality

With perfect precision, the boundary conditions of the minimum partition balance (the hyperbolic figure eight knot) define the limits of measure in physical reality—the limits of time, space, charge, mass and temperature known as the Planck constants. Each external rotation (ϕ_k) participating in the decomposition balance of the hyperbolic figure eight knot multiplied by a double cover ($2\boldsymbol{n}$) raised to the number of derangements defining each rotation (d_k), defines a dimensional boundary of physical reality: $boundary_k = \phi_k(2\boldsymbol{n})^{\pm d_k}$.

Planck temperature $T_P = \phi_4(2\boldsymbol{n})^{+d_4}$

$T_P = 1.4167869859079 \ldots \times 10^{32}$ predicted
$T_P = 1.416784(16) \times 10^{32}\ K$ measured

Planck mass $m_P = \phi_3(2\boldsymbol{n})^{-d_3}$

$m_P = 2.1764268381757 \ldots \times 10^{-8}$ predicted
$m_P = 2.176434(24) \times 10^{-8}\ kg$ measured

Planck charge $q_P = \phi_2(2\boldsymbol{n})^{-d_2}$

$q_P = 1.8755459671396 \ldots \times 10^{-18}$ predicted
$q_P = 1.8755459 \times 10^{-18}\ C$ previously defined

Planck length $l_P = \phi_1(2\boldsymbol{n})^{-d_1}$

$l_P = 1.6162591817564 \ldots \times 10^{-35}$ predicted
$l_P = 1.616255(18) \times 10^{-35}\ m$ measured

Planck time $t_P = \phi_0(2\boldsymbol{n})^{-d_0}$

$t_P = 5.3912583683231 \ldots \times (10)^{-44}$ predicted
$t_P = 5.391247(60) \times (10)^{-44}\ s$ measured

Where $(t_P, l_P, q_P, m_P, T_P)$ = the Planck time, length, charge, mass, and temperature, $(\phi_0, \phi_1, \phi_2, \phi_3, \phi_4)$ = the 5 partition rotations of the hyperbolic figure eight knot, $(d_0, d_1, d_2, d_3, d_4)$ = the number of derangements defining each of those rotations.

The simplest self-balanced geometry intrinsically maintains the *coherent system of units* that frame reality. The partitioned balance of the hyperbolic figure eight knot defines the boundaries of reality's 5 fundamental dimensions—time, space, charge, mass, and temperature.

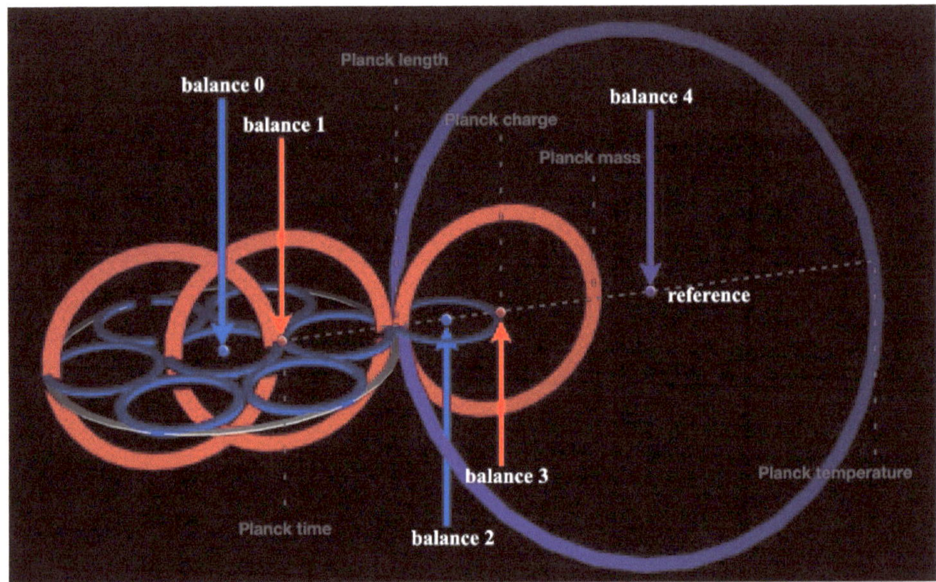

The Planck constants are defined as the action limit (maximum reach/boundary) of each balance: t_P *= the maximum reach of balance 0,* l_P *= the maximum reach of balance 1,* q_P *= the maximum reach of balance 2,* m_P *= the maximum reach of balance 3,* T_P *= the maximum reach of boundary 4.*

Let's examine how the external boundaries of this balanced projection connect.

Chapter 6: the hyperbolic vortex

The hyperbolic figure eight knot's exterior domain is maintained under self-closed hyperbolic connection.

$$\frac{1}{ж} + ж + ж^3 = \text{hyperbolic connection}$$

$$\frac{1}{ж} + ж + \frac{ж^3}{2\pi} = \text{self} - \text{closed hyperbolic connection}$$

Where ж is the variable $0 \leq ж \leq 1$.

The action responsible for this self-closed hyperbolic connection trivially involves 2 orthogonal twists, both of which are internal to the mass boundary.

$$\left(i^i \right)^{-\frac{\pi}{2}} - m_P$$

Where $i^i = e^{-\frac{\pi}{2}}$ is a single twist (a rotation of $\frac{\pi}{2}$ radians), $\left(i^i \right)^{-\frac{\pi}{2}}$ is an orthogonal combination of 2 internal twists, and $m_P =$ the Planck mass boundary.

Setting these two conditions equal to each other yields the hyperbolic vortex equation.

the hyperbolic vortex equation

$$\frac{1}{ж} + ж + \frac{ж^3}{2\pi} = \left(i^i \right)^{-\frac{\pi}{2}} - m_P$$

Where $\left(i^i \right)^{-\frac{\pi}{2}} =$ 2 internal orthogonal twists of $\frac{\pi}{2}$, and $m_P =$ the Planck mass.

The hyperbolic vortex equation has 4 solutions composed of 6 parts (4 real and 2 imaginary).

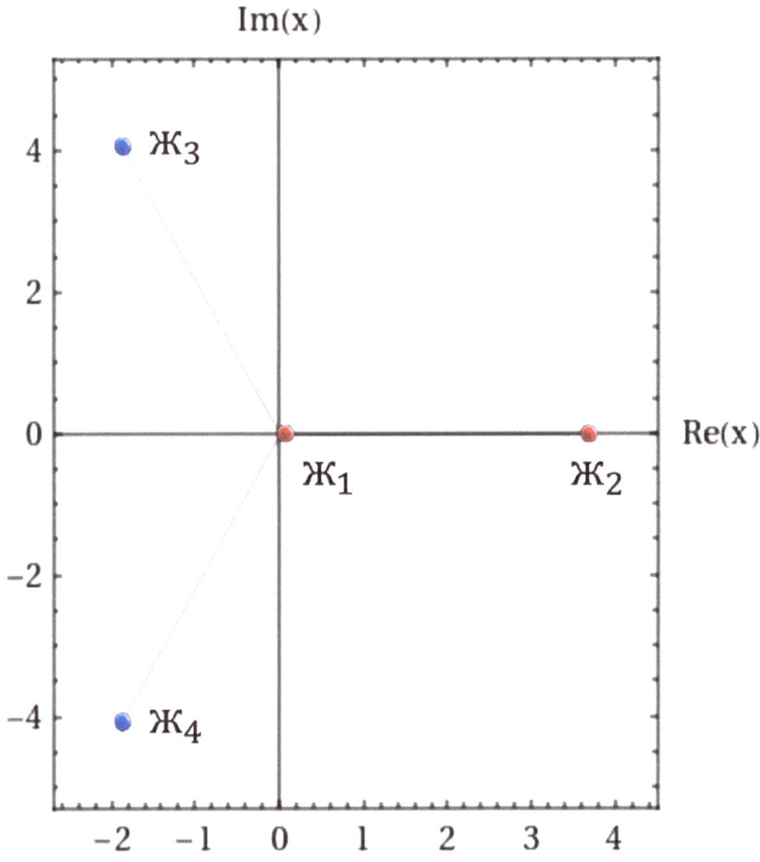

The 4 hyperbolic vortex partition constants.

$$ж_1 = 0.0854245431533310\ldots$$
$$ж_2 = 3.66756753485499\ldots$$
$$ж_3 = -1.87649603900416\ldots + 4.06615262615971\ldots i$$
$$ж_4 = -1.87649603900416\ldots - 4.06615262615971\ldots i$$

These hyperbolic vortex partition constants are ideally cyclic and combinatorial. For example, their product is equal to 2π. They sum to zero, as do their real and imaginary parts. Their squares sum to the projection of a sphere (-4π). And so on.

$ж_1ж_2ж_3ж_4 = 2\pi$ product

$ж_1 + ж_2 + ж_3 + ж_4 = 0$ complex zero sum

$ж_1 + ж_2 + Re(ж_3) + Re(ж_4) = 0$ real zero sum

$Im(ж_3) + Im(ж_4) = 0$ imaginary zero sum

${ж_1}^2 + {ж_2}^2 + {ж_3}^2 + {ж_4}^2 = -4\pi$ square sum

$Re(ж_3) = Re(ж_4)$ twin

$Im(ж_3) = -Im(ж_4)$ reflection

$\left(Re(ж_3)\right)^2 + \left(Im(ж_3)\right)^2 = \left(Re(ж_4)\right)^2 + \left(Im(ж_4)\right)^2 = ж_3ж_4$ pair

$\left(Re(ж_3)\right)^2 + \left(Im(ж_4)\right)^2 = \left(Re(ж_4)\right)^2 + \left(Im(ж_3)\right)^2 = ж_3ж_4$ pair

$\left(\frac{{ж_1}^2 + {ж_2}^2 + {ж_3}^2 + {ж_4}^2}{ж_1 \times ж_2 \times ж_3 \times ж_4}\right) = -2$ self intersection number

$ж_3$ and $ж_4$ can also be expressed in terms of radius and angle.

$ж_r = 4.47826244916751 \ldots$ hyperbolic vortex radius constant
$ж_\theta = 2.00316562310924 \ldots$ hyperbolic vortex radian constant

$ж_r = \sqrt{ж_3ж_4}$

$ж_\theta = \tan^{-1}\left(\frac{Re(ж_3)}{Im(ж_4)}\right) + \frac{\pi}{2}$

${ж_r}^2 = ж_3ж_4$

$ж_1ж_2{ж_r}^2 = 2\pi$

These partitions solutions define how the external facing boundaries of the hyperbolic figure eight knot divide under closed hyperbolic connection. Therefore, they define how the Planck *charge* and Planck *mass* boundaries naturally partition. Let's explore those partitions.

$$\sum_{k=1}^{4} ж_k = 0 \qquad \prod_{k=1}^{4} ж_k = 2\pi$$

$$\sum_{k=1}^{4} {ж_k}^2 = -4\pi \qquad \prod_{k=1}^{4} {ж_k}^2 = 4\pi^2$$

$$-\pi \sum_{k=1}^{4} {ж_k}^2 = \prod_{k=1}^{4} {ж_k}^2$$

hyperbolic vortex partition relation

$$\left(\prod_{k=1}^{4} ж_k\right) \div \left(\sum_{k=1}^{4} {ж_k}^2\right) = -\frac{1}{2}$$

product to square sum

$$\frac{ж_1 ж_2 ж_3 ж_4}{4} = \frac{\pi}{2}$$

multiplicative average

Combinatorics of the hyperbolic vortex partition constants.

Chapter 7: charge partitions

Externally, *charge* has a magnitude equal to the charge boundary (the Planck charge q_P) multiplied by the 1st hyperbolic vortex partition constant ($ж_1$), defining the charge of the electron. (see Chapter 10).

$$e = ж_1 q_P$$

Internally *charge* partitions exactly as the volume of the hyperbolic figure eight knot partitions in the imaginary (internal) direction, into a negative one-thirds and a positive two-thirds division (of a dilogarithmic flip $Li_2(-1)$).

$$V_{fe} = i\left[-Li_2\left((-1)^{\frac{1}{3}}\right) + Li_2\left((-1)^{\frac{2}{3}}\right)\right]$$

Where Li_2 is the hyperbolic dilogarithm.

This arranges *charge* into the following fundamental partitions.

charge partitions

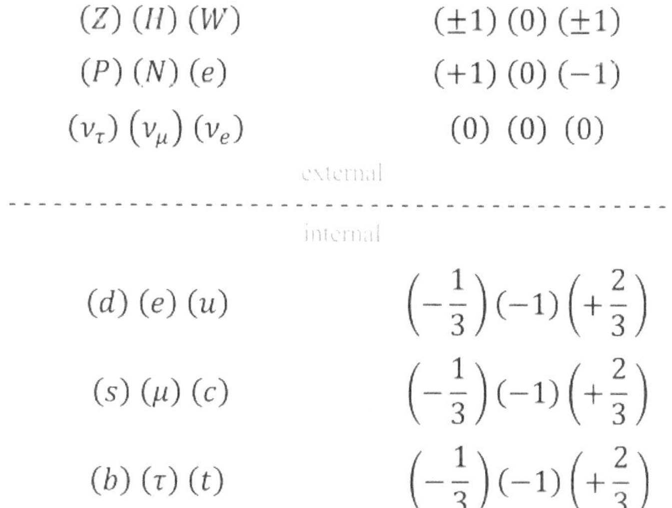

$(Z)\ (H)\ (W)$	$(\pm 1)\ (0)\ (\pm 1)$
$(P)\ (N)\ (e)$	$(+1)\ (0)\ (-1)$
$(\nu_\tau)\ (\nu_\mu)\ (\nu_e)$	$(0)\ (0)\ (0)$
external	
internal	
$(d)\ (e)\ (u)$	$\left(-\frac{1}{3}\right)(-1)\left(+\frac{2}{3}\right)$
$(s)\ (\mu)\ (c)$	$\left(-\frac{1}{3}\right)(-1)\left(+\frac{2}{3}\right)$
$(b)\ (\tau)\ (t)$	$\left(-\frac{1}{3}\right)(-1)\left(+\frac{2}{3}\right)$

Where $e = ж_1 q_P$ is set as the scale of charge unity $= 1$, and Z, H, W, P, N, e, ν_τ, ν_μ, ν_e, d, u, s, μ, c, b, τ and t are the charge assignments of the Z boson, Higgs boson, W boson, proton, neutron, electron, tau neutrino, muon neutrino, electron neutrino, down quark, up quark, strange quark, muon, charm quark, beauty (bottom) quark, tau quark, and truth (top) quark.

Chapter 8: mass partitions

The mass boundary (m_P) has a geometry that trivially binds 2 figure eight knot volumes (V_{fe}) in 4 orthogonal ways and terminates inverse-hyperbolically over the derangement structure of the hyperbolic figure eight knot. This primitive tessellation of the Planck mass boundary defines the mass of the electron (m_e).

$$m_e = 2V_{fe}\, m_P{}^4 \left(1 + \left(\sinh\left(\sinh\left(\frac{!n}{b}\right)\right)^{-1}\right)^{-1} ⊞\right)$$

$m_e = 9.10938370161994\ldots \times 10^{-31}\, kg$ predicted

$m_e = 9.1093837015(28) \times 10^{-31}\, kg$ measured

Where V_{fe} = the internal complement volume of the hyperbolic figure eight knot, m_P = the Planck mass boundary, $\sinh(x)$ = the hyperbolic sine function, $n = 5$ the number of unique rotations partitioning the hyperbolic figure eight knot, $!n = 44$ the number of derangements available to 5 things, $b = 7$ the break in scale symmetry internally maintained by the balance of the hyperbolic figure eight knot, and $⊞ = \left(\frac{l_P\, m_P}{q_P{}^2}\right)$ is the termination boundary of the hyperbolic figure eight knot—the split squared intersection of the Planck length, mass, and charge (l_P, m_P, q_P) boundaries.

The external action of *every* inverted cyclic balance maintained by this ideal geometry terminates on this internal/external division boundary $⊞ = \left(\frac{l_P\, m_P}{q_P{}^2}\right)$, where the sum of the hyperbolic figure eight knot's 5 rotations externally partition into the n-hypersphere of *maximal* volume.

$$\sum_{k=0}^{4} \phi_k = \left(\frac{2}{3}\right)^2 \frac{4\pi}{Re(\omega_1)^3}\left(1 + \left(\frac{3}{2}\right)^2 \left(\frac{V_h^*}{n}\right) ⊞\right)$$

Where $\sum \phi_k$ = the sum of the unique rotations partitioning the hyperbolic figure eight knot, $Re(\omega_1)$ = the Real part of the omega_1 constant (the equianharmonic half-period from Chapter 4), V_h^* = the dimension at which an n-hypersphere has *maximal* volume, and ⊞ = the termination boundary.

Any squarely balanced (periodic) hyperbolic space has a total of 17 possible tessellations, or tilings. (There are exactly 17 plane symmetry groups.) To find all of the possible tessellations of the Planck *mass* boundary we orthogonally extend its trivial tessellation (m_e) under ideal balance of (splitting)^2 and joining—guaranteeing that:

1. the Planck charge and Planck mass boundaries partition under hyperbolic vortex connection ($ж_1, ж_2, \ldots$), and
2. the balance of that connection squarely hyperbolically bifurcates (splits) in 2 orthogonal but otherwise indistinguishable ways $(\alpha_F - 1)^2$, while
3. dilogarithmically persisting under a unitary hyperbolic-circular closed balance (being logarithmically (μ, γ) braided under the $+\left(\frac{2}{3}\right)$ and $-\left(\frac{1}{3}\right)$ internal split action of the hyperbolic figure eight knot).

The tessellations that capture this precise balance of orthogonal *square splitting* and *joining* define the proton mass (m_+) and the neutron mass (m_N).

$$\left(\frac{m_e}{m_+}\right)\left(\frac{ж_2}{ж_1}\right)^2 = \left(\frac{2}{3}\right)^2 (\alpha_F - 1)^2 \left(1 - \left(\frac{1}{3}\right) e^{3\gamma} ⊞\right) \quad \text{splitting map}$$

$$\left(\frac{m_N - m_+}{m_e}\right) = \left(\frac{1}{3}\right)(\mu + 3 + \pi)\left(1 + \left(\frac{2}{3}\right) e^{3\gamma} ⊞\right) \quad \text{joining map}$$

$m_+ = 1.67262192371195 \ldots \times 10^{-27}\, kg$ predicted
$m_+ = 1.67262192369(51) \times 10^{-27}\, kg$ measured

$m_N = 1.67492749802284 \ldots \times 10^{-27}\, kg$ predicted
$m_N = 1.67492749804(95) \times 10^{-27}\, kg$ measured

Where m_+ = the proton mass, m_N = the neutron mass, α_F = the alpha Feigenbaum constant (the 2nd period-doubling bifurcation constant of the logistic map), μ = the nontrivial zero of the logarithmic integral γ = the Euler-Mascheroni constant (the limiting difference between the harmonic series and the natural logarithm function), $ж_1$ & $ж_2$ = the 1st and 2nd hyperbolic vortex partition constants, and ⊞ = the termination boundary.

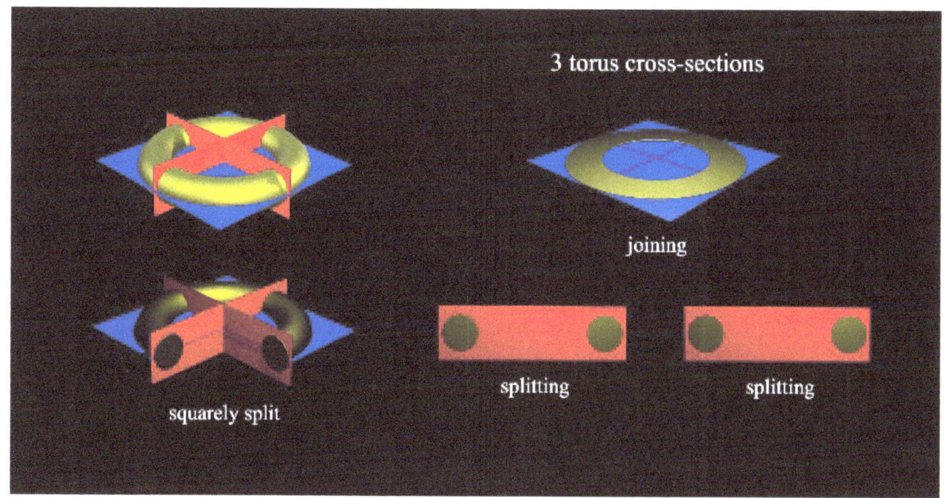

The action of the hyperbolic figure eight knot externally fibers under 2 orthogonal circular actions (whose product space is the asymmetric Clifford torus), defining the orthogonal intersection of 2 squarely arranged circle-splittings and 1 joining of circles.

Mirroring the division structure of the hyperbolic figure eight knot's internal complement volume, which is composed of 6 circles surrounding a central circle that successively undergoes 4 orthogonal internal bifurcations,

$$
\begin{aligned}
&= \left(\frac{m_H - m_Z}{m_W}\right)6 \\
&= \left(\frac{m_N - m_+}{m_e}\right) \\
&= \left(\frac{m_d - m_u}{m_e}\right)\left(\frac{1}{2}\right) \\
&= \left(\frac{m_c - 2m_s}{m_\mu}\right)\left(\frac{1}{2}\right)^2 \\
&= \left(\frac{m_t - m_\tau}{m_b}\right)\left(\left(\frac{1}{2}\right)^2\right)^2 \\
&= \left(\frac{m_{\nu_\tau} - m_{\nu_\mu}}{m_{\nu_e}}\right)\left(\left(\left(\frac{1}{2}\right)^2\right)^2\right)^2
\end{aligned}
$$

the joining map of these 3 trivial tessellations (m_e, m_+, m_N) extends into a balance of externally phased and internally folded tessellations. And the orthogonally connected splitting map extends into a balance of twists, splits and flips. Both of these orthogonal extensions define all 17 mass tilings.

full joining map

$$\left(\frac{m_H - m_Z}{m_W}\right) = \left(\frac{1}{3}\right)^2 (\mu + 3 + \pi)(2^{-1})\left(1 + \left(\frac{\text{ж}_2{}^2}{\text{ж}_1{}^3}\right) G_{Ga}{}^{-3}\left(\frac{n}{b}\right) \boxplus\right)$$

$$\left(\frac{m_N - m_+}{m_e}\right) = \left(\frac{1}{3}\right) (\mu + 3 + \pi)(2^{0})\left(1 + \left(\frac{2}{3}\right) e^{3\gamma} \boxplus\right)$$

$$\left(\frac{m_d - m_u}{m_e}\right) = \left(\frac{1}{3}\right) (\mu + 3 + \pi)(2^{+1})\left(1 + \left(\frac{\text{ж}_2{}^2}{\text{ж}_1{}^3}\right) D_{Do}{}^{3}\sqrt{\frac{2}{b}} \boxplus\right)$$

external phases

---

internal folds

$$\left(\frac{m_c - 2m_s}{m_\mu}\right) = \left(\frac{1}{3}\right)(\mu + 3 + \pi)(2^{2})\left(1 + 0 \;\; \boxplus\right)$$

$$\left(\frac{m_t - m_\tau}{m_b}\right) = \left(\frac{1}{3}\right)(\mu + 3 + \pi)\left(2^{2^{2}}\right)\left(1 + b\frac{\sqrt{Re(\omega_1)}}{\text{ж}_1{}^4} \boxplus\right)$$

$$\left(\frac{m_{\nu_\tau} - m_{\nu_\mu}}{m_{\nu_e}}\right) = \left(\frac{1}{3}\right)(\mu + 3 + \pi)\left(2^{2^{2^{2}}}\right)\left(1 + 0 \;\; \boxplus\right)$$

Where m_H, m_Z, m_W, m_N, m_+, m_e, m_d, m_u, m_c, m_s, m_μ, m_t, m_τ, m_b, m_{ν_τ}, m_{ν_μ}, and m_{ν_e} are the masses of the Higgs boson, Z boson, W boson, neutron, proton, electron, down quark, up quark, charm quark, strange quark, muon, truth (top) quark, tau quark, beauty (bottom) quark, tau neutrino, muon neutrino, and the electron neutrino; μ = the Ramanujan-Soldner constant (the nontrivial zero of the logarithmic integral), γ = the Euler-Mascheroni constant (which defines the limiting difference between the natural logarithm and the harmonic series), ж_1 and ж_2 = the 1st and 2nd hyperbolic vortex partition constants, G_{Ga} = Gauss's constant, D_{Do} = the unique real root of the cosine function (a universal attracting fixed point called the Dottie number), ω_1 = the omega_1 constant, and $\boxplus$ = the termination boundary of the hyperbolic figure eight knot.

full splitting map

$$\frac{m_b + m_c + m_t}{\left(\sqrt{m_b} + \sqrt{m_c} + \sqrt{m_t}\,\right)^2} = \left(\frac{2}{3}\right)^{3} (\,\alpha_F - 1\,)^{2} \left(1 + \left(\frac{ж_2{}^3}{ж_1{}^2}\right) \frac{V_{fe}{}^4}{2^{n}}\ ⊞\right)$$

$$\left(\frac{m_e}{m_+}\right)\left(\frac{ж_2}{ж_1}\right)^2 = \left(\frac{2}{3}\right)^{2} (\,\alpha_F - 1\,)^{2} \left(1 - \left(\frac{1}{3}\right) e^{3\gamma}\ ⊞\right)$$

$$\frac{m_e + m_\mu + m_\tau}{\left(\sqrt{m_e} + \sqrt{m_\mu} + \sqrt{m_\tau}\,\right)^2} = \left(\frac{2}{3}\right)^{1} (\,\alpha_F - 1\,)^{0} \left(1 - (\,3\,ж_2{}^2\,)\ P_{up}\quad ⊞\right)$$

- -

$$\frac{m_H + m_Z + m_W}{\left(\sqrt{m_H} + \sqrt{m_Z} + \sqrt{m_W}\,\right)^2} = \left(\frac{2}{3}\right)^{-\frac{1}{2}} (\alpha_F - 1)^{-e^{2\gamma}} \left(1 + \left(\frac{ж_2}{ж_1}\right)^2 \left(2\boldsymbol{nb}\,(V_{fe})^{\frac{1}{3}}\right)^{-\frac{1}{2}}\ ⊞\right)$$

$$\frac{m_u + m_s + m_d}{\left(\sqrt{m_u} + \sqrt{m_s} + \sqrt{m_d}\,\right)^2} = \left(\frac{2}{3}\right)^{0} \gamma\,(\,\alpha_F - 1\,)^{0} \left(1 + 0\ \ ⊞\right)$$

$$\frac{m_{\nu_e} + m_{\nu_\mu} + m_{\nu_\tau}}{\left(\sqrt{m_{\nu_e}} + \sqrt{m_{\nu_\mu}} + \sqrt{m_{\nu_\tau}}\,\right)^2} = \left(\frac{2}{3}\right)^{-\frac{1}{2}} (\alpha_F - 1)^{+e^{\pi i}} \left(1 - \left(\frac{ж_2}{ж_1}\right)^2 \left(2\boldsymbol{nb}\,(V_{fe})^{\frac{2}{3}}\right)^{-\frac{1}{2}}\ ⊞\right)$$

Where m_b, m_c, m_t, m_e, m_+, m_μ, m_τ, m_H, m_Z, m_W, m_u, m_s, m_d, m_{ν_e}, m_{ν_μ} and m_{ν_τ} are respectively the masses of the: beauty quark, charm quark, truth quark, electron, proton, muon, tau quark, Higgs boson, Z boson, W boson, up quark, strange quark, down quark, electron neutrino, muon neutrino, and tau neutrino, α_F = the alpha Feigenbaum constant, γ = the Euler-Mascheroni constant (which denotes the limiting difference between the natural logarithm and the harmonic series), $ж_1$ and $ж_2$ = the 1st and 2nd hyperbolic vortex partition constants, V_{fe} = the internal volume complement of the hyperbolic figure eight knot, P_{up} = the universal parabolic constant, and ⊞ = the termination boundary of the hyperbolic figure eight knot.

partition parameters of the hyperbolic figure eight knot

$n = 5$ number of unique rotations
$b = 7$ break in scale symmetry

$\phi_0 = 5.39125836832313 \ldots + 2\pi i(k) \quad k \in \mathbb{Z}$ 0th external rotation constant
$\phi_1 = 1.61625918175645 \ldots + 2\pi i(k)$ 1st external rotation constant
$\phi_2 = 1.87554596713962 \ldots + 2\pi i(k)$ 2nd external rotation constant
$\phi_3 = 2.17642683817579 \ldots + 2\pi i(k)$ 3rd external rotation constant
$\phi_4 = 1.41678698590795 \ldots + 2\pi i(k)$ 4th external rotation constant

$t_P = 5.39125836832313 \ldots \times 10^{-44}\, s$ Planck time
$l_P = 1.61625918175645 \ldots \times 10^{-35}\, m$ Planck length
$q_P = 1.87554596713962 \ldots \times 10^{-18}\, C$ Planck charge
$m_P = 2.17642683817579 \ldots \times 10^{-8}\, kg$ Planck mass
$T_P = 1.41678698590795 \ldots \times 10^{32}\, K$ Planck temperature

$G_{Gi} = 1.01494160640965 \ldots$ Gieseking's constant
$V_{fe} = 2.02988321281930 \ldots$ figure eight knot complement volume
$e = 2.71828182845904 \ldots$ Euler's number
$\pi = 3.14159265358979 \ldots$ Archimedes' constant
$ж_1 = 0.0854245431533304 \ldots$ 1st hyperbolic vortex partition constant
$ж_2 = 3.66756753485499 \ldots$ 2nd hyperbolic vortex partition constant
$ж_3 = -1.87649603900417 \ldots + 4.06615262615972 \ldots i$ 3rd hvpc
$ж_4 = -1.87649603900417 \ldots - 4.06615262615972 \ldots i$ 4th hvpc
$ж_r = 4.47826244916751 \ldots$ hyperbolic vortex radius constant
$ж_\theta = 2.00316562310924 \ldots$ hyperbolic vortex radian constant
$\alpha_F = 2.50290787509589 \ldots$ alpha Feigenbaum constant
$\delta_F = 4.66920160910299 \ldots$ delta Feigenbaum constant
$\gamma = 0.577215664901532 \ldots$ Euler-Mascheroni constant
$\mu = 1.45136923488338 \ldots$ nontrivial zero of the logarithmic integral
$G_{Ga} = 0.834626841674073 \ldots$ Gauss's constant
$L = 2.622057554292119 \ldots$ lemniscate constant
$L_1 = 1.31102877714605 \ldots$ 1st lemniscate constant
$L_2 = 0.599070117367796 \ldots$ 2nd lemniscate constant
$D_{Do} = 0.739085133215160 \ldots$ Dottie number
$\omega_1 = 0.764977018528596 \ldots + 1.32497062714087 \ldots i$ omega_1
$\omega_2 = 1.529954037057192 \ldots$ omega_2 constant
$P_{up} = 2.29558714939263 \ldots$ universal parabolic constant
$W_{We} = 0.474949379987920 \ldots$ Weierstrass constant
$A_h^* = 7.25694640486057 \ldots$ dimension of maximal n-hypersphere area
$V_h^* = 5.25694640486057 \ldots$ dimension of maximal n-hypersphere volume

the 17 tessellations of the Planck mass boundary

$m_e = 9.10938370161994 \ldots \times 10^{-31}\ kg$ electron mass

$m_+ = 1.67262192371195 \ldots \times 10^{-27}\ kg$ proton mass

$m_N = 1.67492749802284 \ldots \times 10^{-27}\ kg$ neutron mass

$m_c = 2.27188026398178 \ldots \times 10^{-27}\ kg$ charm quark mass

$m_d = 8.44242715614137 \ldots \times 10^{-30}\ kg$ down quark mass

$m_u = 3.81810683898335 \ldots \times 10^{-30}\ kg$ up quark mass

$m_s = 1.82501207639326 \ldots \times 10^{-28}\ kg$ strange quark mass

$m_b = 7.45149186313980 \ldots \times 10^{-27}\ kg$ beauty (bottom) quark mass

$m_t = 3.08390948667753 \ldots \times 10^{-25}\ kg$ truth (top) quark mass

$m_H = 2.23150010999262 \ldots \times 10^{-25}\ kg$ Higgs boson mass

$m_Z = 1.62556627846185 \ldots \times 10^{-25}\ kg$ Z boson mass

$m_W = 1.43263881046217 \ldots \times 10^{-25}\ kg$ W boson mass

$m_\tau = 3.16754001786349 \ldots \times 10^{-27}\ kg$ tau mass

$m_\mu = 1.88353162775445 \ldots \times 10^{-28}\ kg$ muon mass

$m_{\nu_\tau} = 7.63391385156818 \ldots \times 10^{-39}\ kg$ tau neutrino mass

$m_{\nu_\mu} = 1.61737751049693 \ldots \times 10^{-43}\ kg$ muon neutrino mass

$m_{\nu_e} = 1.07825167366462 \ldots \times 10^{-43}\ kg$ electron neutrino mass

Where the black digits represent previously known values (either measured or geometrically known), green digits represent extended predictions, and red digits represent discrepancies between prediction and measurement. Note, there are no red digits.

The 17 tessellations of the Planck mass boundary precisely define the mass values of the 17 fundamental particles of matter.

orthogonal maps of the Planck mass boundary

$$m_e = 2V_{fe}\, m_P{}^4 \left(1 + \left(\sinh\left(\sinh\left(\frac{!n}{b}\right)\right)^{-1}\right)^{-1} \boxplus\right) \quad \text{balance map}$$

$$\left(\frac{m_e}{m_+}\right)\left(\frac{ж_2}{ж_1}\right)^2 = \left(\frac{2}{3}\right)^2 (\alpha_F - 1)^2 \left(1 - \left(\frac{1}{3}\right) e^{3\gamma} \boxplus\right) \quad \text{splitting map}$$

$$\left(\frac{m_N - m_+}{m_e}\right) = \left(\frac{1}{3}\right)(\mu + 3 + \pi)\left(1 + \left(\frac{2}{3}\right) e^{3\gamma} \boxplus\right) \quad \text{joining map}$$

Chapter 9: individual tessellations

Every unique tessellation of the mass boundary represents a fundamental geometric tiling from which the hyperbolic figure eight knot's external *mass boundary* can be periodically constructed. These tilings are either geometrically combinatorial (defining a balance between 2 actions, the second of which ends on the termination boundary), or primitive (referencing a single division action).

Let's examine the geometry of each tiling.

charm quark mass

$$\frac{m_e}{m_c}\left(\frac{ж_2{}^2}{ж_1{}^2}\right) = D_{DO}$$

Where m_e = the electron mass, and D_{DO} = the Dottie number, the unique real fixed point of the cosine function—a universal attracting point

$m_c = 2.27188026398178 \ldots \times 10^{-27}\ kg$ predicted
$m_c = 2.272(63) \times 10^{-27}\ kg$ measured

down quark mass

$$\frac{m_e}{m_d}(ж_2{}^2) = \mu$$

Where μ = the nontrivial zero of the logarithmic integral

$m_d = 8.44242715614137 \ldots \times 10^{-30}\ kg$ predicted
$m_d = 8.3(07) \times 10^{-30}\ kg$ measured

up quark mass

$$\frac{m_e}{m_u} = \frac{2}{\omega_2{}^n}$$

Where ω_2 = the omega_2 constant

$$\omega_2 = \frac{\Gamma^3\left(\frac{1}{3}\right)}{4\pi}$$

$m_u = 3.81810683898335 \ldots \times 10^{-30}\ kg$ predicted
$m_u = 3.92(89) \times 10^{-30}\ kg$ measured

strange quark mass

$$\frac{m_e}{m_s}(ж_2{}^2) = \frac{\sqrt{n\,K_{\_1}}}{!\,n}$$

Where $K_{\_1}$ = the Khinchin harmonic mean

$m_s = 1.82501207639326 \ldots \times 10^{-28}\ kg$ predicted
$m_s = 1.69(16) \times 10^{-28}\ kg$ measured

bottom quark mass

$$\frac{m_e}{m_b}\left(\frac{ж_2{}^2}{ж_1{}^2}\right) = \frac{L_1{}^3}{2n}$$

Where L_1 = the 1st lemniscate constant

$m_b = 7.45149186313980 \ldots \times 10^{-27}\ kg$ predicted
$m_b = 7.45(07) \times 10^{-27}\ kg$ measured

top quark mass

$$\frac{m_e}{m_t}\left(\frac{ж_2{}^3}{ж_1{}^2}\right) = \frac{L_2}{6n}$$

Where L_2 = the 2nd lemniscate constant

$m_t = 3.08390948667753 \ldots \times 10^{-25}\ kg$ predicted
$m_t = 3.084(07) \times 10^{-25}\ kg$ measured

Higgs mass

$$\frac{m_e}{m_H}\left(\frac{ж_2{}^2}{ж_1{}^2}\right) = \frac{b}{2\, j_{0,1}{}^b}$$

Where $j_{0,1}$ = the 1st root of the Bessel function

$m_H = 2.23150010999262 \ldots \times 10^{-25}\ kg$ predicted
$m_H = 2.2315(28) \times 10^{-25}\ kg$ measured

Z boson mass

$$\frac{m_e}{m_Z}\left(\frac{ж_2{}^3}{ж_1{}^2}\right) = \left(\frac{2}{3}\right)^3 L_{LL}{}^n$$

Where L_{LL} = the Laplace limit, $\frac{1}{L_{LL}} = sinh(\,C_{CFP}\,)$, and C_{CFP} = the real fixed point of the hyperbolic cotangent

$m_Z = 1.62556627846185 \ldots \times 10^{-25}\ kg$ predicted
$m_Z = 1.625566(38) \times 10^{-25}\ kg$ measured

W boson mass

$$\frac{m_e}{m_W}\left(\frac{ж_2{}^2}{ж_1{}^3}\right) = \left(Im(\rho_1)\right)^{-3/4}$$

Where ρ_1 = the first nontrivial root of the zeta function

$m_W = 1.43263881046217 \ldots \times 10^{-25}\ kg$ predicted
$m_W = 1.43288(21) \times 10^{-25}\ kg$ measured

tau mass

$$\frac{m_e}{m_\tau}\left(\frac{ж_2{}^2}{ж_1{}^2}\right) = \left(\frac{3}{V_{fe}{}^2}\right)^2 \left(1 - ж_2{}^2\, sech\left(tan\left(\frac{1}{2}\right)\right) ⊞\right)$$

$m_\tau = 3.16754001786349 \ldots \times 10^{-27}\ kg$ predicted
$m_\tau = 3.16754(21) \times 10^{-27}\ kg$ measured

muon mass

$$\frac{m_e}{m_\mu}\left(\frac{1}{ж_1{}^2}\right) = L_{LL}\left(1 + \frac{\boldsymbol{b}\, ж_2}{V_{fe}{}^2}\left(\frac{!\boldsymbol{n}}{\sqrt{\boldsymbol{n}}}\right) ⊞\right)$$

Where L_{LL} = the Laplace limit, $\frac{1}{L_{LL}} = sinh(\,C_{CFP}\,)$, C_{CFP} = the real fixed point of the hyperbolic cotangent, and ⊞ = the termination boundary of the hyperbolic figure eight knot

$m_\mu = 1.883531627754459 \ldots \times 10^{-28}\ kg$ predicted
$m_\mu = 1.883531627(42) \times 10^{-28}\ kg$ measured

tau neutrino mass

$$\frac{\phi_1(2\boldsymbol{n})^{-\boldsymbol{bn}}}{m_{\nu_\tau}}\left(\frac{ж_1}{ж_2}\right)^2 = 2^{1/\boldsymbol{n}}\left(1 - (\sqrt{2} - 1)\left(\frac{ж_2}{ж_1}\right)^2 ⊞\right)$$

$m_{\nu_\tau} = 7.63391385156818 \ldots \times 10^{-39}\ kg$ predicted
$m_{\nu_\tau} = ???\ kg$ no current measurement

muon neutrino mass

$$m_{\nu_\mu} = 3\,\phi_0 (2\boldsymbol{n})^{-!\boldsymbol{n}}$$

$m_{\nu_\mu} = 1.61737751049693 \ldots \times 10^{-43}\ kg$ predicted

$m_{\nu_\mu} = ???\ kg$ no current measurement

electron neutrino mass

$$m_{\nu_e} = 2\,\phi_0 (2\boldsymbol{n})^{-!\boldsymbol{n}}$$

$m_{\nu_e} = 1.07825167366462 \ldots \times 10^{-43}\ kg$ predicted

$m_{\nu_e} = ???\ kg$ no current measurement

Chapter 10: constants of Nature

So far, we have found that the balance of the hyperbolic figure eight knot is responsible for the persistent arena of reality, minimally decomposing into 5 unique actions (time, space, charge, mass and temperature) with sharply defined boundaries (the Planck constants). And that the external domain of that balance connects via the hyperbolic vortex equation, which partitions the Planck charge and Planck mass boundaries into the exact charge and mass values that define the fundamental particles of matter.

In this chapter, we discover that every unique balance of boundaries maintaining this minimum geometry defines a constant of Nature.

To unveil the geometry of each constant of Nature, we think of them as binomial combinations of a primary partition action A_p on a primary intersection of boundaries B_p, and a terminal partition action A_t on the terminal boundary intersection ⊞.

universal binomial factorization prescription

$$(constant\ of\ Nature) = A_p B_p (1 \pm A_t\ ⊞)$$

Since the terminal action doesn't come into play until the 7th digit, this equation can be solved in pieces. That is, since $⊞ = \left(\frac{l_P\, m_P}{q_P{}^2}\right) = 1 \times 10^{-7}$ we can safely ignore the terminal contribution of this balance at first, solve for its primary action and primary boundary $A_p B_p$, and then return to the full equation to solve for the terminal action A_t.

Step 1

Temporarily set $A_t = 0$, and construct the primary boundary B_p from the dimensions of the constant, in terms of the natural partitions we have defined so far $(t_P, l_P, q_P, m_P, T_P, m_e, m_+, \ldots)$.

Step 2

If $A_p \neq 1$, then solve for it in terms of the hyperbolic figure eight knot's partition parameters $(ж_1, ж_2, ж_r, \pi, G_{Gi}, V_{fe}, e, \gamma, \alpha_F, \mathbf{b}, \mathbf{n} \ldots)$.

Step 3

Plug $A_p B_p$ into the universal binomial factorization prescription and solve for the terminal geometric action A_t of that constant of Nature.

Let's work through 2 examples.

speed of light $$c = 2.99792458 \times 10^8 \frac{m}{s}$$

To find the primary boundary intersection of any constant of Nature we look at its balance of dimensions. For example, since the speed of light c has dimensions *meters/seconds*, its primary boundary B_p will be equal to the natural boundary of length (the Planck length) divided by the natural boundary of time (the Planck time).

$$B_p = \frac{l_P}{t_P} = 2.99792566287110 \ldots \times 10^8\, m/s$$

Since this number already yields the measured value for the speed of light up to the first 7 digits, we know that for this constant $A_p = 1$. To finish, we plug $A_p B_p$ back into the universal binomial factorization prescription and solve for the terminal action A_t of the speed of light.

$$c = \frac{l_P}{t_P}(1 - A_t \boxplus)$$

Finding that,

$$c = \frac{l_P}{t_P}\left(1 - \left(\frac{3}{2}\right) j_{0,1} \boxplus\right)$$

Where $j_{0,1}$ = the 1st root of the Bessel function.

quantized Hall conductance $$H_C = 3.87404614(17) \times 10^{-5} \frac{s\, C^2}{m^2 kg}$$

Following the same procedure for the quantized Hall conductance H_C we ignore its terminal contribution and match the dimensions of this constant to its appropriate Planck boundaries. Since its dimensions are seconds Coulombs squared divided by meters squared kilograms $s\, C^2/m^2 kg$, we set its primary boundary equal to the Planck time multiplied by the Planck charge squared divided by the Planck length squared and the Planck mass.

$$B_p = \frac{t_P\, q_P{}^2}{l_P{}^2\, m_P}$$

Next, we solve for the primary action in terms of the hyperbolic figure eight knot's partition parameters: $A_p = \frac{ж_1{}^2}{2\pi}$. Then we plug $A_p B_p$ back into the universal binomial factorization prescription and solve for the terminal partition action of this balance: $A_t = ж_2$.

$$H_C = \frac{ж_1{}^2}{2\pi}\left(\frac{t_P\, q_P{}^2}{l_P{}^2 m_P}\right)\left(\, 1 - ж_2 \boxplus \,\right)$$

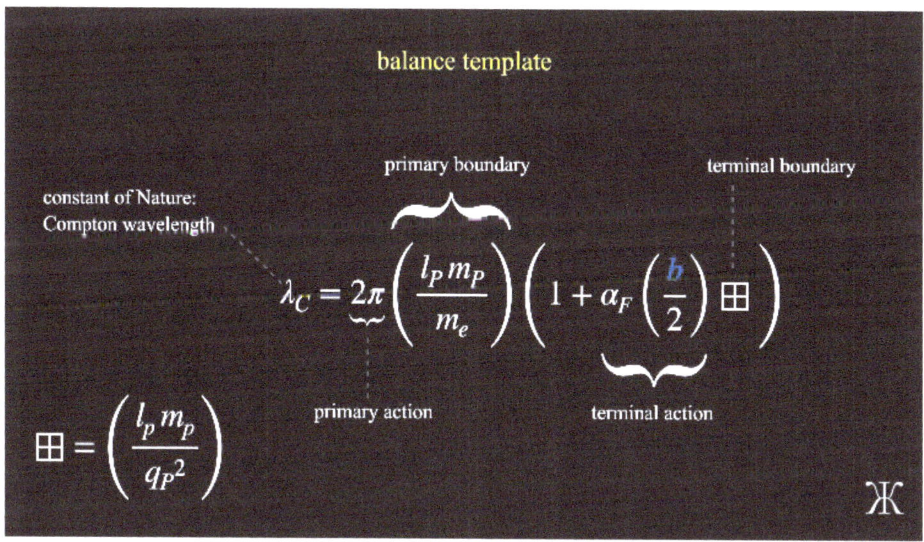

Every inverted balance is a binomial union, joining a primary geometric action on a primary boundary (the construction of which is determined by the dimensions of the constant) with a terminal geometric action on the fixed terminal boundary.

This universal binomial factorization prescription allows us to identify the constructive geometry of every constant of Nature. [4]

---

[4] Note that 1 mol combines the dimensions of $q_P\, m_P$.

fine-structure constant

$$\alpha = ж_1{}^2$$

$\alpha = 7.29735257295522\ldots \times 10^{-3}$ predicted
$\alpha = 7.2973525698(24) \times 10^{-3}$ measured

elementary charge

$$e = ж_1 q_P$$

$e = 1.60217657405973\ldots \times 10^{-19}\ C$ predicted
$e = 1.602176565(35) \times 10^{-19}\ C$ measured

Compton wavelength

$$\lambda_C = 2\pi\left(\frac{l_P\, m_P}{m_e}\right)\left(1 + \alpha_F\left(\frac{b}{2}\right)⊞\right)$$

$\lambda_C = 2.42631023893941\ldots \times 10^{-12}\ m$ predicted
$\lambda_C = 2.4263102389(16) \times 10^{-12}\ m$ measured

Josephson constant

$$K_J = \frac{ж_1}{\pi}\left(\frac{t_P\, q_P}{l_P{}^2\, m_P}\right)\left(1 - \left(\frac{e^{2\gamma}}{Im(\omega_1)^3}\right)\left(\frac{b}{2}\right)⊞\right)$$

Where $\gamma =$ the Euler-Mascheroni constant, and $\omega_1 =$ the omega_1 constant

$K_J = 4.835978484 00467\ldots \times 10^{14}\ \ s\,C/m^2kg$ predicted
$K_J = 4.835978484 \times 10^{14}\ \ s\,C/m^2kg$ previously defined

Planck's constant

$$\hbar = \left(\frac{l_P{}^2\, m_P}{t_P}\right)\left(1 + \sqrt{\frac{b\pi}{\zeta(3)}}\ ⊞\right)$$

Where $\zeta(3) =$ Apéry's constant

$\hbar = 1.05457172593010\ldots \times 10^{-34}\ \ m^2kg/s$ predicted
$\hbar = 1.054571726(47) \times 10^{-34}\ \ m^2kg/s$ measured

electric constant

$$\varepsilon_0 = \frac{1}{4\pi}\left(\frac{t_P{}^2\, q_P{}^2}{l_P{}^3\, m_P}\right)\left(1 - \left(\frac{n^2}{2^n}\right)\boxplus\right)$$

$\varepsilon_0 = 8.85418781308692 \ldots \times 10^{-12}\ s^2C^2/m^3kg$ predicted
$\varepsilon_0 = 8.8541878128(13) \times 10^{-12}\ s^2C^2/m^3kg$ measured

Coulomb's constant

$$\kappa = \left(\frac{l_P{}^3\, m_P}{t_P{}^2\, q_P{}^2}\right)\left(1 + \left(\frac{n^2}{2^n}\right)\boxplus\right)$$

$\kappa = 8.98755179196986 \ldots \times 10^9\ m^3kg/s^2C^2$ exact
$\kappa = 8.9875517923(14) \times 10^9\ m^3kg/s^2C^2$ measured

quantized Hall conductance

$$H_C = \frac{ж_1{}^2}{2\pi}\left(\frac{t_P\, q_P{}^2}{l_P{}^2 m_P}\right)\left(1 - ж_2 \boxplus\right)$$

Where $ж_1 ж_2 ж_3 ж_4 = ж_1 ж_2 ж_r{}^2 = 2\pi$, and $ж_1{}^2 = \alpha$ the fine structure constant, and $\boxplus$ = the termination boundary

$H_C = 3.87404614816855 \times 10^{-5}\, s\, C^2/m^2kg$ predicted
$H_C = 3.87404614(17) \times 10^{-5}\, s\, C^2/m^2kg$ measured

Bohr magneton

$$\mu_B = \frac{ж_1}{2}\left(\frac{l_P{}^2\, q_P\, m_P}{t_P\, m_e}\right)\left(1 + \frac{2}{ж_2}\pi^2 \boxplus\right)$$

Where $ж_1 ж_2 ж_3 ж_4 = ж_1 ж_2 ж_r{}^2 = 2\pi$

$\mu_B = 9.27400994938886 \ldots \times 10^{-24}\ m^2C/s$ predicted
$\mu_B = 9.274009994(57) \times 10^{-24}\ m^2C/s$ measured

magnetic constant

$$\mu_0 = 4\pi \boxplus \left(1 + ж_2\, e^{\gamma} \sqrt{\frac{3}{2}} \boxplus \right)$$

Where γ = the Euler-Mascheroni constant

$\mu_0 = 1.25663706143747 \ldots \times 10^{-6}\ m\,kg/C^2$ predicted
$\mu_0 = 1.256637061 \ldots \times 10^{-6}\ m\,kg/C^2$ previously defined

1st radiation constant

$$c_1 = 4\pi^2 \left(\frac{l_P{}^4\, m_P}{t_P{}^3} \right) \left(1 - ж_2{}^2 (\, 2\gamma - 1\,) \boxplus \right)$$

$c_1 = 3.74177185217629 \ldots \times 10^{-16}\ m^4 kg/s^3$ predicted
$c_1 = 3.741771852 \ldots \times 10^{-16}\ m^4 kg/s^3$ previously defined

electron Thomson cross section

$$\sigma_e = \left(\frac{2}{3} \right) 4\pi\, ж_1{}^4 \left(\frac{l_P\, m_P}{m_e} \right)^2 \left(1 + ж_2{}^2 (\, L - 1\,) \boxplus \right)$$

Where L = the lemniscate constant, and $ж_1 ж_2 ж_r{}^2 = 2\pi$

$\sigma_e = 6.65246159951664 \ldots \times 10^{-29}\ m^2$ predicted
$\sigma_e = 6.6524616(18) \times 10^{-29}\ m^2$ measured

conductance quantum

$$G_0 = \frac{ж_1{}^2}{\pi} \left(\frac{t_P\, q_P{}^2}{l_P{}^2 m_P} \right) \left(1 - \frac{ж_2{}^2}{\mu^3} \boxplus \right)$$

Where μ = the nontrivial zero of the logarithmic integral, $ж_1{}^2 = \alpha$ the fine structure constant, and $\boxplus$ = the termination boundary

$G_0 = 7.74809172907834 \ldots \times 10^{-5}\ s\, C^2/m^2 kg$ predicted
$G_0 = 7.748091729 \ldots \times 10^{-5}\ s\, C^2/m^2 kg$ previously defined

Bohr electron radius

$$a_0 = \frac{1}{ж_1{}^2}\left(\frac{l_P\, m_P}{m_e}\right)\left(1 + ж_2{}^2\left(\frac{6}{Im(\rho_1)}\right)^{1/2} ⊞\right)$$

Where $Im(\rho_1)$ = the imaginary part of the first nontrivial root of the zeta function, and $ж_1, ж_2$ are the 1st and 2nd hyperbolic vortex partition constants

$a_0 = 5.29177210936601 \ldots \times 10^{-11}\ m$ predicted
$a_0 = 5.2917721092(17) \times 10^{-11}\ m$ measured

atomic mass constant

$$m_u = m_e\left(\frac{ж_2}{ж_1}\right)^2 G_{Gi}{}^{-\frac{3}{4}}\left(1 - \frac{sinh\left(csc\left(\frac{4}{3}n\right)\right)}{ж_r{}^2} ⊞\right)$$

Where m_e = the electron mass, G_{Gi} = Gieseking's constant, $ж_1$ and $ж_2$ = the 1st and 2nd hyperbolic vortex partition constants, $ж_r$ = the hyperbolic vortex radius constant, and $ж_1 ж_2 ж_r{}^2 = 2\pi$

$m_u = 1.66053906659995 \ldots \times 10^{-27}\ kg$ predicted
$m_u = 1.66053906660(50) \times 10^{-27}\ kg$ measured

characteristic impedance

$$Z_0 = 4\pi\left(\frac{l_P{}^2 m_P}{t_P\, q_P{}^2}\right)\left(1 + \frac{2n^2}{b}\Gamma(x_{min})^4 ⊞\right)$$

Where $\Gamma(x_{min})$ = the minimal positive argument value of the gamma function, and ⊞ = the termination boundary

$Z_0 = 3.76730313668332 \ldots \times 10^2\ m^2kg/s\ C^2$ predicted
$Z_0 = 3.76730313668(57) \times 10^2\ m^2kg/s\ C^2$ measured

Stefan-Boltzmann constant

$$\sigma = \frac{\zeta(2)}{2n}\left(\frac{m_P}{t_P{}^3\,T_P{}^4}\right)\left(1 + \left(\frac{P_{up}}{2\,ж_1}\right)^2 \boxplus\right)$$

Where P_{up} = the universal parabolic constant

$\sigma = 5.67037441935166 \ldots \times 10^{-8}\ kg/s^3K^4$ predicted
$\sigma = 5.670374419 \times 10^{-8}\ kg/s^3K^4$ previously defined

Avogadro constant

$$N_A = \frac{6\,ж_1{}^2}{e^\gamma}\left(\frac{1}{q_P\,m_P}\right)\left(1 - \left(2\,ж_2\,P_{up}\right)^2 \boxplus\right)$$

$N_A = 6.02214076693260 \ldots \times 10^{23}\ 1/mol$ predicted
$N_A = 6.02214076 \times 10^{23}\ 1/mol$ previously defined

von Klitzing constant

$$R_K = \frac{2\pi}{ж_1{}^2}\left(\frac{l_P{}^2\,m_P}{t_P\,q_P{}^2}\right)\left(1 + \frac{1}{2}Re\left(i^{i^{i^{\cdots}}}\right)ж_r{}^2 \boxplus\right)$$

Where $Re\left(i^{i^{i^{\cdots}}}\right)$ = the real part of the infinite power tower of the imaginary root i, $ж_1{}^2 = \alpha$ the fine structure constant, and $ж_r$ = the hyperbolic vortex radius constant, and $ж_1 ж_2 ж_r{}^2 = 2\pi$

$R_K = 2.5812807449400712 \ldots \times 10^4\ m^2kg/s\,C^2$ predicted
$R_K = 2.58128074434(84) \times 10^4\ m^2kg/s\,C^2$ measured

Hartree energy

$$E_h = ж_1{}^4\left(\frac{l_P{}^2\,m_e}{t_P{}^2}\right)\left(1 - ж_2\,\omega_2{}^4\left(\frac{3}{n}\right)^2 \boxplus\right)$$

Where ω_2 = the omega_2 constant

$E_h = 4.3597447222118 3 \ldots \times 10^{-18}\ m^2kg/s^2$ predicted
$E_h = 4.3597447222071(85) \times 10^{-18}\ m^2kg/s^2$ measured

the speed of light

$$c = \left(\frac{l_P}{t_P}\right)\left(1 - \left(\frac{3}{2}\right) j_{0,1} ⊞\right)$$

Where $j_{0,1}$ = the 1st root of the Bessel function

$c = 2.99792458144786 \ldots \times 10^8\ m/s$ predicted
$c = 2.99792458 \times 10^8\ m/s$ previously defined

spectral radiance constant

$$c_{1L} = 4\pi \left(\frac{{l_P}^4\, m_P}{{t_P}^3}\right)\left(1 - \sqrt{\frac{3}{2}}\, j_{0,1} ⊞\right)$$

$c_{1L} = 1.19104286900535 \ldots \times 10^{-16}\ m^4 kg/s^3$ predicted
$c_{1L} = 1.191042869(53) \times 10^{-16}\ m^4 kg/s^3$ measured

classical electron radius

$$r_e = ж_1{}^2 \left(\frac{l_P\, m_P}{m_e}\right)\left(1 + n\,(\alpha_F - 1)\sqrt{\left(\frac{2}{3}\right) V_{fe}}\ ⊞\right)$$

Where α_F = the alpha Feigenbaum constant, and V_{fe} = the complement volume of the hyperbolic figure eight knot

$r_e = 2.817940322269510 \ldots \times 10^{-15}\ m$ predicted
$r_e = 2.8179403227(19) \times 10^{-15}\ m$ measured

Nuclear magneton

$$\mu_N = \frac{ж_1}{2}\left(\frac{{l_P}^2\, q_P\, m_P}{t_P\, m_+}\right)\left(1 + ж_r{}^2 \left(\frac{2}{3}\right)\frac{V_{fe}}{n} ⊞\right)$$

$\mu_N = 5.05078369897026 \ldots \times 10^{-27}\ m^2 C/s$ predicted
$\mu_N = 5.050783699(31) \times 10^{-27}\ m^2 C/s$ measured

2nd radiation constant

$$c_2 = 2\pi\,(\, l_P\, T_P\,)\left(\, 1 - 2\boldsymbol{n}\, e^{2\gamma}\, C_{CFP}{}^2 \boxplus \,\right)$$

Where C_{CFP} = the real fixed point of the hyperbolic cotangent

$c_2 = 1.43877687708189 \ldots \times 10^{-2}\ m\,K$ predicted
$c_2 = 1.438776877 \ldots \times 10^{-2}\ m\,K$ previously defined

muon g-factor

$$g_\mu = -\frac{1}{4}(\,\alpha_F - 1\,)\sqrt{\frac{V_{fe}{}^{\boldsymbol{b}}}{\boldsymbol{n}}}\left(1 + \frac{1}{\boldsymbol{b}}\sqrt{\frac{\boldsymbol{n}}{\boldsymbol{b}}} \boxplus\right)$$

Where α_F = the alpha Feigenbaum constant, and V_{fe} = the complement volume of the hyperbolic figure eight knot

$g_\mu = -2.00233184179953 \ldots$ predicted
$g_\mu = -2.00233184122(82)$ measured

electron g-factor

$$g_e = -\frac{1}{4}(\,\alpha_F - 1\,)\sqrt{\frac{V_{fe}{}^{\boldsymbol{b}}}{\boldsymbol{n}}}\left(1 - \left(\frac{ж_2{}^2}{ж_1{}^3}\right)\frac{1}{2\boldsymbol{n}^2}\left(\frac{3}{F_{FR}{}^2}\right)^2 \boxplus\right)$$

Where F_{FR} = the Fransén-Robinson constant

$g_e = -2.00231930436231 \ldots$ predicted
$g_e = -2.00231930436256(35)$ measured

molar gas constant

$$R = \frac{6\,ж_1{}^2}{e^\gamma}\left(\frac{l_P{}^2}{t_P{}^2\, q_P\, T_P}\right)\left(1 - \left(\frac{ж_2{}^2}{ж_1{}^3}\right)\frac{!\boldsymbol{n}}{2\boldsymbol{n}}\left(\frac{3}{F_{FR}{}^3}\right)^3 \boxplus\right)$$

$R = 8.31446261831660 \ldots\ m^2 kg/s^2 K\ mol$ predicted
$R = 8.314462618\ m^2 kg/s^2 K\ mol$ previously defined

magnetic flux constant

$$\Phi_0 = \left(\frac{\pi}{ж_1}\right)\left(\frac{l_P{}^2\, m_P}{t_P\, q_P}\right)\left(1 + 8\, G_{Go} \boxplus\right)$$

Where G_{Go} = the Gompertz constant

$\Phi_0 = 2.06783384804388 \ldots \times 10^{-15}\, m^2 kg/s\, C$ predicted
$\Phi_0 = 2.067833848 \ldots \times 10^{-15}\, m^2 kg/s\, C$ previously defined

quantum of circulation

$$q_c = \pi\left(\frac{l_P{}^2\, m_P}{t_P\, m_e}\right)\left(1 + 8\, C_C \boxplus\right)$$

Where C_C = Cahen's constant

$q_c = 3.63694755171206 \ldots \times 10^{-4}\, m^2/s$ predicted
$q_c = 3.6369475516(11) \times 10^{-4}\, m^2/s$ measured

proton g-factor

$$g_+ = ж_2\,(\,2\,B_1 + 1\,)\left(1 + 3\, C_c{}^{-4} \boxplus\right)$$

Where B_1 = Merten's constant and C_C = Cahen's constant

$g_+ = +5.58569468933155 \ldots$ predicted
$g_+ = +5.5856946893(16)$ measured

Rydberg constant

$$R_\infty = \frac{ж_1{}^4}{4\pi}\left(\frac{m_e}{l_P\, m_P}\right)\left(1 - 4\, L_2\, b^{2/3} \boxplus\right)$$

Where L_2 = the 2nd lemniscate constant, $ж_1{}^2 = \alpha$ the fine structure constant, and $\boxplus$ = the termination boundary

$R_\infty = 1.09737315685870 \ldots \times 10^7\ 1/m$ predicted
$R_\infty = 1.0973731568539(55) \times 10^7\ 1/m$ measured

proton gyromagnetic ratio

$$\gamma_+ = \frac{\pi}{n} csch^2\left(\frac{1}{2}\right)^2 \left(\frac{ж_2}{ж_1}\right) ж_r{}^2 \left(\frac{t_P}{q_P\, m_e}\right)\left(1 + \left(\frac{ж_2}{ж_1}\right)\frac{1}{G'} ⊞\right)$$

Where the Wilbraham-Gibbs constant $G' = Si(\pi)$ the sine integral of π

$\gamma_+ = 2.67522187458888 \ldots \times 10^8\ s/kg\ C$ predicted
$\gamma_+ = 2.6752218744(11) \times 10^8\ s/kg\ C$ measured

Neutron magnetic moment

$$N_\mu = -\left(\frac{ж_1}{2}\right)\left(\frac{!n}{\Gamma(n) - 1}\right)\left(\frac{l_P{}^2 q_P\, m_P}{t_P\, m_+}\right)$$

Where $!x$ = the derangement function, $\Gamma(x)$ = the gamma function, and $\Gamma(x+1) = x!$ = the factorial function.

$N_\mu = -9.66236357094966 \ldots \times 10^{-27}\ m^2C/s$ predicted
$N_\mu = -9.6623647(23) \times 10^{-27}\ m^2C/s$ measured

Faraday constant

$$F = N_A\, ж_1\, q_P \left(1 + \frac{ж_r{}^2}{2\pi}\left(\frac{n}{!n}\right) ⊞\right)$$

Where N_A = Avogadro's number, $ж_1 q_P = e$ the electron charge and $ж_r$ = the hyperbolic vortex radius constant ($ж_1 ж_2 ж_r{}^2 = 2\pi$)

$F = 9.64853321242908 \ldots \times 10^4\ \ C/mol$ predicted
$F = 9.648533212 \ldots \times 10^4\ \ C/mol$ previously defined

neutron g-factor

$$g_N = -\left(\frac{ж_\theta}{n}\right)\left(3G_{Gi}{}^2\right)^2\left(1 + \Gamma(n) ⊞\right)$$

Where G_{Gi} = Gieseking's constant

$g_N = -3.82608515049597 \ldots$ predicted
$g_N = -3.82608545(90)$ measured

gravitational coupling

$$\alpha_G = \left(\frac{m_e}{m_P}\right)^2$$

$\alpha_G = 1.75182147492404 \ldots \times 10^{-45}$ predicted
$\alpha_G = 1.7518(21) \times 10^{-45}$ measured

Compton angular frequency

$$\omega_C = \left(\frac{m_e}{t_P\, m_P}\right)\left(1 - \frac{1}{2}\Gamma(\boldsymbol{n}) \boxplus\right)$$

$\omega_c = 7.76344099944556 \ldots \times 10^{20}\ 1/s$ predicted
$\omega_c = 7.763441 \times 10^{20}\ 1/s$ measured

Schwinger magnetic induction

$$S_{mi} = \frac{1}{ж_1}\left(\frac{{m_e}^2}{t_P\, q_P\, m_P}\right)\left(1 - 2\pi\left(3\, Im\left(i^{i^{i\cdots}}\right)\right)^2 \boxplus\right)$$

$S_{mi} = 4.41899541452692 \ldots \times 10^{9}\ kg/s\ C$ predicted
$S_{mi} = 4.419 \times 10^{9}\ kg/s\ C$ measured

gravitational constant

$$G = \left(\frac{{l_P}^3}{{t_P}^2\, m_P}\right)\left(1 - \left(\frac{1}{2}\pi\, {ж_r}^2\right)\Gamma(\boldsymbol{n}) \boxplus\right)$$

$G = 6.67384038951738 \ldots \times 10^{-11}\ m^3/s^2kg$ predicted
$G = 6.67384(80) \times 10^{-11}\ m^3/s^2kg$ measured

Boltzmann constant

$$k_B = \left(\frac{{l_P}^2\, m_P}{{t_P}^2\, T_P}\right)\left(1 - \left(\frac{1}{2}\pi\, ж_r\right)^2 \boxplus\right)$$

$k_B = 1.38064931695409 \ldots \times 10^{-23}\ m^2kg/s^2K$ predicted
$k_B = 1.380649 \times 10^{-23}\ m^2kg/s^2K$ previously defined

proton radius

$$\frac{r_e}{r_+} = L\sqrt{\frac{3}{2}}$$

Where r_e = the classical electron radius, and L = the lemniscate constant

$r_+ = 8.7749356797017 \times 10^{-16}\ m$ predicted
$r_+ = 8.751(61) \times 10^{-16}\ m$ measured

The most mysterious parameters in existence are no longer the parameters *of* existence. The simplest possible partition balance is responsible for all the action parameters of quantum field theory and general relativity. This minimal balance persistently maintains the 5 Planck boundaries, and the connections between these boundaries trivially define the 4 fundamental actions of Nature: the strong force, the weak force, the electromagnetic force, and gravity.

The strength of the gravitational force is determined by the ratio of the Planck length cubed to the Planck time squared and the Planck mass, terminally arranged under trivial spherical factorization. And the strength of the electromagnetic force compared to the strong nuclear force is equal to the 1st hyperbolic vortex partition constant squared.

$$G = \left(\frac{l_P{}^3}{t_P{}^2\, m_P}\right)\left(1 - \left(\frac{1}{2}\pi\, ж_r{}^2\right)\Gamma(\boldsymbol{n})\,⊞\right) \qquad \frac{em\ force}{strong\ force} = ж_1{}^2$$

If we square that ratio again, we get the ratio of the classical electron radius compared to the Bohr electron radius. And the radius of the Helium atom, and the Lithium atom, and so on[5] are:

$$\frac{r_e}{a_0} = ж_1{}^4 \qquad He^+ = \frac{a_0}{2} \qquad Li^{2+} = \frac{a_0}{3} \qquad \ldots$$

---

[5] The nonrelativistic ground state wave function of the hydrogen atom is

$$\psi(r) = \frac{a_0{}^{-\frac{3}{2}}}{\Gamma\left(\frac{1}{2}\right)}\ e^{-\frac{r}{a_0}}$$

Where $\Gamma(x)$ = the gamma function, and a_0 = the Bohr electron radius.

the constants of Nature

$\alpha = 7.29735257295522 \ldots \times 10^{-3}$	fine-structure constant
$e = 1.60217657405973 \ldots \times 10^{-19}\ C$	electron charge
$\lambda_C = 2.42631023893941 \ldots \times 10^{-12}\ m$	Compton wavelength
$K_J = 4.83597848400467 \ldots \times 10^{14}\ sC/m^2kg$	Josephson constant
$\hbar = 1.05457172593010 \ldots \times 10^{-34}\ m^2kg/s$	Planck's constant
$\varepsilon_0 = 8.85418781308692 \ldots \times 10^{-12}\ s^2C^2/m^3kg$	electric constant
$\kappa = 8.98755179196986 \ldots \times 10^{9}\ m^3kg/s^2C^2$	Coulomb's constant
$H_C = 3.87404614816855 \times 10^{-5}\ C^2/m^2kg$	quantized Hall conductance
$\mu_B = 9.27400994938886 \ldots \times 10^{-24}\ m^2C/s$	Bohr magneton
$\mu_0 = 1.25663706143747 \ldots \times 10^{-6}\ mkg/C^2$	magnetic constant
$c_1 = 3.74177185217629 \ldots \times 10^{-16}\ m^4kg/s^3$	1st radiation constant
$\sigma_e = 6.65246159951664 \ldots \times 10^{-29}\ m^2$	electron Thomson x section
$G_0 = 7.74809172907834 \ldots \times 10^{-5}\ sC^2/m^2kg$	conductance quantum
$a_0 = 5.29177210936601 \ldots \times 10^{-11}\ m$	Bohr electron radius
$m_u = 1.66053906659995 \ldots \times 10^{-27}\ kg$	atomic mass constant
$Z_0 = 3.76730313668332 \ldots \times 10^{2}\ m^2kg/sC^2$	characteristic impedance
$\sigma = 5.67037441935166 \ldots \times 10^{-8}\ kg/s^3K^4$	Stefan-Boltzmann constant
$N_A = 6.02214076693260 \ldots \times 10^{23}\ 1/mol$	Avogadro constant
$R_K = 2.58128074494007 \ldots \times 10^{4}\ m^2kg/sC^2$	von Klitzing constant
$E_h = 4.35974472220674 \ldots \times 10^{-18}\ m^2kg/s^2$	Hartree energy
$c = 2.99792458144786 \ldots \times 10^{8}\ m/s$	speed of light
$c_{1L} = 1.19104286900535 \ldots \times 10^{-16}\ m^4kg/s^3$	spectral radiance
$r_e = 2.81794032269510 \ldots \times 10^{-15}\ m$	classical electron radius
$\mu_N = 5.05078369897026 \ldots \times 10^{-27}\ m^2C/s$	Nuclear magneton
$c_2 = 1.43877687708189 \ldots \times 10^{-2}\ m\ K$	2nd radiation constant
$g_\mu = -2.00233184179953 \ldots$	muon g-factor
$g_e = -2.00231930436231 \ldots$	electron g-factor
$R = 8.31446261831660 \ldots\ m^2kg/s^2K\ mol$	molar gas constant
$\Phi_0 = 2.06783384804388 \ldots \times 10^{-15}\ m^2kg/sC$	magnetic flux constant
$q_c = 3.63694755171206 \ldots \times 10^{-4}\ m^2/s$	quantum of circulation
$g_+ = +5.58569468933155 \ldots$	proton g-factor
$R_\infty = 1.09737315685870 \ldots \times 10^{7}\ 1/m$	Rydberg constant
$\gamma_+ = 2.67522187458888 \ldots \times 10^{8}\ s/kg\ C$	proton gyromagnetic ratio
$N_\mu = -9.66236357094966 \ldots \times 10^{-27}\ m^2C/s$	neutron magnetic moment
$F = 9.64853321242908 \ldots \times 10^{4}\ C/mol$	Faraday constant
$g_N = -3.82608515049597 \ldots$	neutron g-factor
$\alpha_G = 1.75182147492404 \ldots \times 10^{-45}$	gravitational coupling constant
$\omega_c = 7.76344409944556 \ldots \times 10^{20}\ 1/s$	Compton angular frequency
$S_{mi} = 4.41899541452692 \ldots \times 10^{9}\ kg/s\ C$	Schwinger magnetic induction

$G = 6.67384038951738 \ldots \times 10^{-11}\ m^3/s^2kg$ gravitational constant
$k_B = 1.38064931695409 \ldots \times 10^{-23}\ m^2kg/s^2K$ Boltzmann constant
$r_+ = 8.7749356797017 \times 10^{-16}\ m$ predicted

Where the black digits represent previously known values (either measured or geometrically known), green digits represent extended predictions, and red digits represent discrepancies between prediction and measurement. Note that there are no red digits.

Since first publishing these geometric explanations of the constants of Nature, one of those constants (the magnetic moment of the muon, called the muon g-factor) has been measured to greater accuracy. This latest Fermilab measurement officially puts the number standard model theorists have been using for the muon g-factor at 4.2 standard deviations outside the measured confidence interval. While the number predicted by the division parameters of the hyperbolic figure eight knot lies inside the new experimental average.

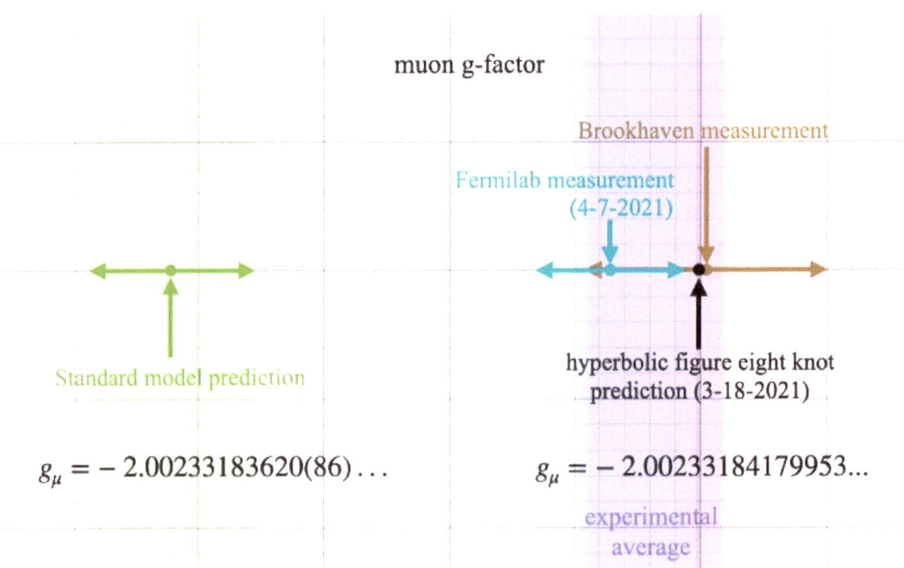

The newest measurement of the muon g-factor supports the hyperbolic figure eight knot.

Chapter 11: Euler's number

So far, we have discovered that the simplest constructable theory of balanced derangements, based on the minimal geometry of the hyperbolic figure eight knot, defines an arena projected under 5 perpetual actions (time, space, charge, mass and temperature) with Planck constant boundaries. The external *charge* and *mass* boundaries of that balance partition into the exact charge and mass values that define the fundamental particles of matter. And every intersection maintained by this minimal balance of boundaries defines a constant of Nature. In other words, the partition parameters of the simplest possible self-balanced geometry precisely define the constructive parameters of physical reality (quantum field theory and general relativity).

In this chapter, we notice that the geometry of this *theory of everything* has a dually hyperbolic union, one that hyperbolically connects *and* hyperbolically partitions. That is, the union of the hyperbolic figure eight knot's bounded actions (the Planck union $U(Planck)$), defined as the sum of the hyperbolic figure eight knot's partitions $\sum p_k$ divided by the product of its rotations $\prod \phi_k$, is equal to the binomial factorized union of the ideal hyperbolic connection (represented by Euler's number e) and the ideal hyperbolic partitioning (represented by the gamma function $\Gamma(x)$).

$$U(Planck) = \frac{\sum p_k}{\prod \phi_k} = e\left(\ 1 + \Gamma(x) ⊞\ \right)$$

Where

$$\sum_{k=0}^{4} p_k = p_0 + p_1 + p_2 + p_3 + p_4 = 137$$

$$\prod_{k=0}^{4} \phi_k = \phi_0 \phi_1 \phi_2 \phi_3 \phi_4 = 50.3938448477126\,...$$

$p_0 = 44$ $\qquad \phi_0 = 5.39125836832313\,...$
$p_1 = 35$ $\qquad \phi_1 = 1.61625918175645\,...$
$p_2 = 18$ $\qquad \phi_2 = 1.87554596713962\,...$
$p_3 = 8$ $\qquad \phi_3 = 2.17642683817579\,...$
$p_4 = 32$ $\qquad \phi_4 = 1.41678698590795\,...$

⊞ = the termination boundary of the hyperbolic figure eight knot, and the argument of the gamma function is set to divide the total derangements of this geometry ($!n$) by the break in scale symmetry of the system (b) and the condition of split symmetry, represented by geometric constant $\bar{s}_{lse} =$ the mean line-between-square edges length (hypercube line picking).[6]

$$x = \frac{!n}{b\,\bar{s}_{lse}} \qquad \bar{s}_{lse} = \left(\frac{1}{3n}\right)\left[\,2+\sqrt{2}+n\,sinh^{-1}(1)\,\right]$$

the Planck union

$$U(Planck) = \frac{\sum p_k}{\prod \phi_k} = e\left(1+\Gamma\left(\frac{!n}{b\,\bar{s}_{lse}}\right)\boxplus\right)$$

Where $\sum p_k =$ the sum of the hyperbolic figure eight knot's unique partition numbers, $\prod \phi_k =$ the product of the hypergolic figure eight knot's unique rotations, $e =$ Euler's number, $\Gamma(x) =$ the gamma function, $!n =$ the derangements of 5 unique rotations, $b =$ the break in scale symmetry of this balance, $\bar{s}_{lse} =$ the mean line-between-square edges length (hypercube line picking), and $\boxplus =$ the terminal boundary of the hyperbolic figure eight knot.

Euler's number $e = 2.71828182845904\ldots$ defines the ideal hyperbolic connection because (1) it defines the base of the *hyperbolic* logarithm, commonly known as the *natural* logarithm, (2) its factorization is ideally harmonic and invertible, and (3) it is *the* number that hyperbolically generalizes. In other words, e is equal to the infinite sum of inverse factorizations, while its multiplicative inverse is equal to the *alternating* infinite sum of inverse factorizations.

$$e = \frac{1}{0!}+\frac{1}{1!}+\frac{1}{2!}+\frac{1}{3!}+\frac{1}{4!}+\cdots = \sum_{k=0}^{\infty}\left(\frac{1}{k!}\right)$$

$$\frac{1}{e} = \frac{1}{0!}-\frac{1}{1!}+\frac{1}{2!}-\frac{1}{3!}+\frac{1}{4!}-\cdots = \sum_{k=0}^{\infty}\left(\frac{(-1)^k}{k!}\right)$$

---

[6] The sum of partitions $\sum p_k = 137 = (\,4^2 + 11^2)$ is equal to the squared sum of the unique factors in our base derangement $!n = 44 = (\,4 \times 11\,)$.

By generalizing this base exponentiation to any factor x,

$$e^x = \frac{x^0}{0!} + \frac{x^1}{1!} + \frac{x^2}{2!} + \frac{x^3}{3!} + \frac{x^4}{4!} \dots = \sum_{k=0}^{\infty} \left(\frac{x^k}{k!} \right)$$

and splitting that generalization into its even and odd parts, we get the hyperbolic identities.

$$cosh(x) = \frac{x^0}{0!} + \frac{x^2}{2!} + \frac{x^4}{4!} + \frac{x^6}{6!} + \frac{x^8}{8!} \dots \qquad cosh(x) = \frac{1}{2}(e^x + e^{-x})$$

$$sinh(x) = \frac{x^1}{1!} + \frac{x^3}{3!} + \frac{x^5}{5!} + \frac{x^7}{7!} + \frac{x^9}{9!} \dots \qquad sinh(x) = \frac{1}{2}(e^x - e^{-x})$$

Where the hyperbolic cosine $cosh(x)$ is the average of the exponential e^x and inverse exponential e^{-x} functions.

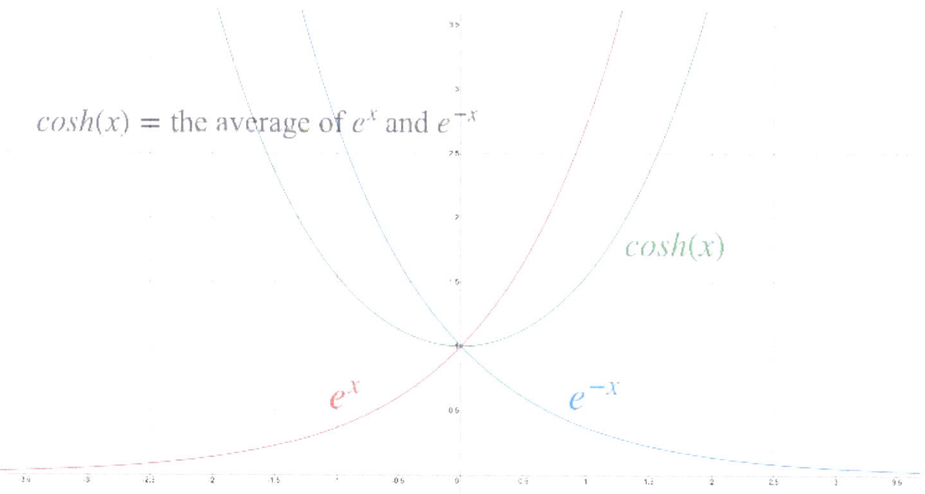

The hyperbolic base of logarithms and the hyperbolic cosine.

And the square of the hyperbolic sine has a golden ratio intersection with the hyperbolic cosine.

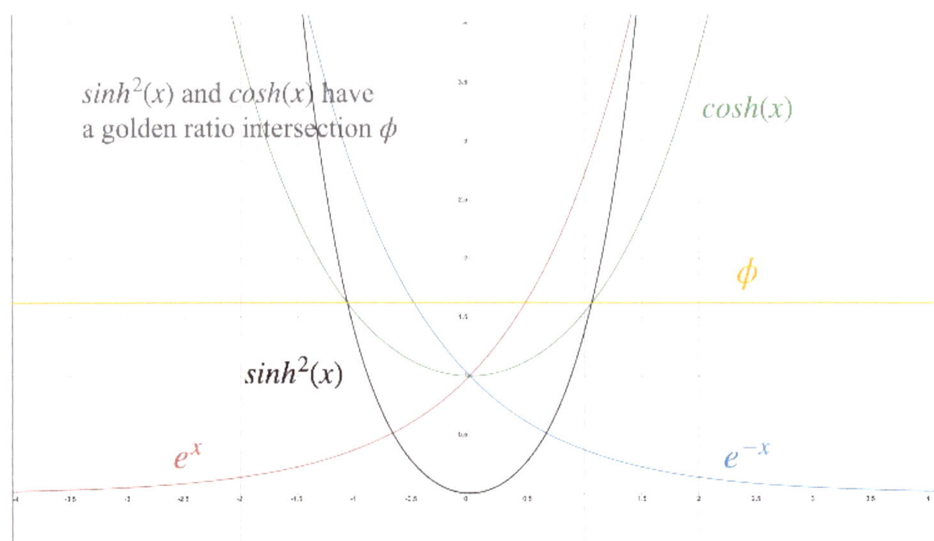

The hyperbolic base of logarithms, the hyperbolic cosine, the hyperbolic sine squared, and the golden ratio.

Now that we've seen how Euler's number *e* defines the ideal hyperbolic connection, let's take a look at the function defining how the terminal part of this balance ideally hyperbolically factors—the gamma function.

Chapter 12: the gamma function

In this chapter, we make the delightful discoverly that the gamma function (the elementary *factorization* function) maps the partition balance of the hyperbolic figure eight knot.

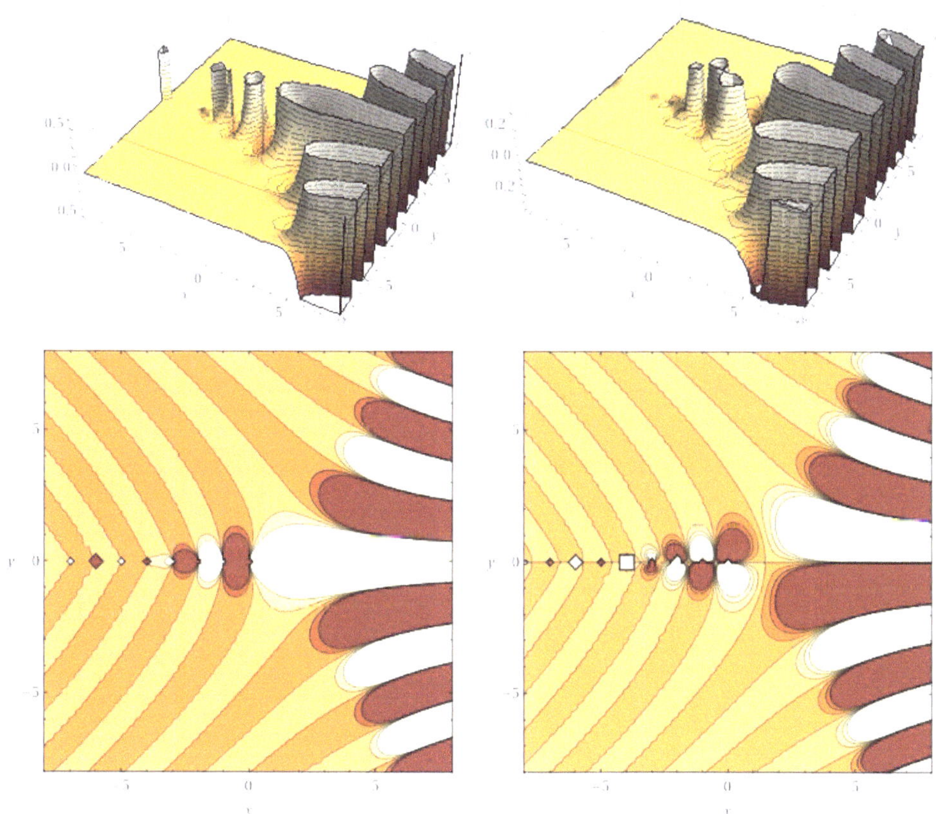

Graph 11: real (left) and imaginary (right) plots of the gamma function under complex argument $\Gamma(x+iy)$.

By symmetry, any successful theory of derangements $(!\,n)$ must also be a successful theory of factorizations $(n!)$.

$$!\,n = n! \sum_{k=0}^{n} \frac{(-1)^k}{k!}$$

Definition of the derangement function $(!\,n)$ *in terms of the factorial function* $(n!)$.

Therefore, the theory of everything that we just laid out in terms of hyperbolically balance derangements must also be simultaneously frameable in terms of hyperbolic factorizations, bringing us to the function many luminaries have declared to be “the most beautiful function of all”.

The gamma function is the unique analytic continuation of the factorial function. This means that instead of being defined only at positive integer values, like the “primitive” factorial function, the gamma function is defined everywhere, except for its simple poles at $s = 0, -1, -2, -3 \ldots$

$$\Gamma(s) = \int_0^{\infty} \frac{x^{s-1}}{e^x}\, dx$$

Integral definition of the gamma function. Where $s = x + iy$.

In short, the gamma function is the *complete* factorization (partition) function. Let’s explore some of its known properties.

The gamma function has no zeros. That is, there is no complex number s for which $\Gamma(s) = 0$. Therefore, the reciprocal gamma function $\frac{1}{\Gamma(s)}$ is an entire function, with zeros at $s = 0, -1, -2, -3 \ldots$

The gamma function hyperbolically generalizes, and has a special point, a local minimum x_{min}.

minimum positive argument value of Γ

$x_{min} = 1.461632144968362 \ldots$

minimal value of Γ for positive argument

$\Gamma(x_{min}) = 0.885603194410888 \ldots$

$$\left|\Gamma\left(0 + xi\right)\right|^2 = \frac{\pi}{x\, sinh(x\pi)}$$

$$\left|\Gamma\left(\frac{1}{2} + xi\right)\right|^2 = \frac{\pi}{cosh(x\pi)}$$

$$\left|\Gamma\left(1 + xi\right)\right|^2 = \frac{x\,\pi}{sinh(x\pi)}$$

The gamma function is also the natural tool for calculating the volume or area of an n-dimensional hypersphere, the arc length of the lemniscate (Chapter 4) and the arc lengths of ellipses, which are given by elliptic integrals simply evaluated in terms of the gamma function.

$$\Gamma(0,x) = -Ei(-x) = Shi(x) - Chi(x)$$

Where $\Gamma(0,x) =$ the incomplete gamma function, $Ei(x) =$ the elliptic integral, $Shi(x) =$ the hyperbolic sine integral, and $Chi(x) =$ the hyperbolic cosine integral.

Algebraically, the gamma function is the unique function that simultaneously satisfies the following 3 conditions for all complex numbers $s = (x + iy)$.

$$\Gamma(1) = 1 \qquad \Gamma(s+1) = s\,\Gamma(s) \qquad \lim_{n\to\infty} \frac{\Gamma(s+n)}{\Gamma(n)\,n^s} = 1$$

The gamma function also naturally splits up into 2 regularized 2-dimensional parts $Q(a,s)$ and $P(a,s)$ defined by the following symmetries.

regularized gamma-functions

$$Q(a,s) + P(a,s) = 1$$ sum of regularized Γ functions

$$\frac{d}{ds}Q(a,s) = -\frac{e^{-s}\,s^{a-1}}{\Gamma(a)}$$ equally split derivatives

$$\frac{d}{ds}P(a,s) = +\frac{e^{-s}\,s^{a-1}}{\Gamma(a)}$$ equally split derivatives

$$\frac{d}{ds}Q(a,s) + \frac{d}{ds}P(a,s) = 0$$ derivative balance

The unitary generalized values of these Q and P regularized gamma functions, are equal to the generalized inverse exponentiation function e^{-x} and its geometric reflection $1 - e^{-x}$.

$$Q(1,x) = e^{-x} \qquad P(1,x) = 1 - e^{-x}$$

In other words, the gamma function internally pulls apart into pieces whose unitary generalized actions define the inverse and reflected-inverse actions of the hyperbolic functions. Its parts constructively generalize in opposition and reflected opposition to the hyperbolic base, Euler's number e.

To unveil the dynamic geometry making use of these ideal hyperbolic factorizations, we set the arguments of the $\mathcal{Q}$ regularized gamma function to the rotations of the time and space boundaries $\mathcal{Q}(\phi_0, \phi_1)$ and examine its output in terms of the hyperbolic figure eight knot's universal binomial prescription. What we find is that the $\mathcal{Q}$ regularized gamma function defines a domain under simple division of the minimum 3-manifold (G_{Gi}) terminally maintaining a 3-dimensional hypersphere connection $(3e^{\pi})$ that is structured over a double cover $(2\boldsymbol{n})$ lattice of that minimum manifold.

$\mathcal{Q}$ regularized under the rotations of time and space

$$\mathcal{Q}(\phi_0, \phi_1) = \left(\frac{1}{G_{Gi}}\right)\left(1 - 3e^{\pi}\left(\frac{1}{G_{Gi}}\right)^{2n} \boxplus\right)$$

Where $\mathcal{Q}(a, s) =$ the $\mathcal{Q}$ regularized gamma function, $\phi_0 \,\&\, \phi_1 =$ the rotations of the time and space boundaries, $G_{Gi} =$ Gieseking's constant, $e^{\pi} =$ the volume sum of all even dimensional hyperspheres, and $\boxplus =$ the terminal boundary of the hyperbolic figure eight knot.

To get the other (orthogonal) half of the story, we set the arguments of the P regularized gamma function to the same time and space rotations $P(\phi_0, \phi_1)$ which, under universal binomial construction, defines the internal and external arrangement of the hyperbolic figure eight knot's partition parameters.

***P* regularized under the rotations of time and space**

$$P(\phi_0, \phi_1) = \left(\frac{\boldsymbol{b}}{{\delta_F}^4}\right)\left(1 - ж_2\sqrt{\frac{2}{W_{We}}} \boxplus\right)$$

Where $P(a, s) =$ the P regularized gamma function, $\phi_0 \,\&\, \phi_1 =$ the rotations of the time and space boundaries, $\boldsymbol{b} =$ the break in scale symmetry, $\delta_F =$ the delta Feigenbaum constant or "bifurcation velocity", $ж_2 =$ the second hyperbolic vortex partition constant, $W_{We} =$ the Weierstrass constant, and $\boxplus =$ the terminal boundary of the hyperbolic figure eight knot.

The gamma function can now be geometrically defined in terms of its balance of internal and external rotations (factors). That is, when the arguments of the Q and P regularized gamma functions are set to the 2 internal rotations of the hyperbolic figure eight knot (ϕ_0, ϕ_1), joined under ideal bifurcation (square/square-root) arrangement, they externally express themselves as (are equal to) the gamma function set to the 4th rotation of the hyperbolic figure eight knot divided by its 2nd and 3rd rotations. The terminal geometry of this arrangement defines the minimum non-compact 3-manifold (G_{Gi}), persisting as a projected balance of n rotations maintained over triple doubly-periodic factorization (W_{We}).

the geometry of the gamma function

$$\left(\delta_F\, Q(\phi_0, \phi_1)\right)^2 \sqrt{\left(P(\phi_0, \phi_1)\right)} = \Gamma\left(\frac{\phi_4}{\phi_2\, \phi_3}\right)\left(1 - (\, G_{Gi}\,) \frac{n}{{W_{We}}^3} \boxplus\right)$$

Where $\delta_F =$ the delta Feigenbaum constant, $Q(a, s) =$ the Q regularized gamma function, $P(a, s) =$ the P regularized gamma function, $\Gamma(x) =$ the gamma function, ϕ_0, ϕ_1, ϕ_2, ϕ_3 & ϕ_4, are the magnitudes of the external rotations maintaining the partition balance of the hyperbolic figure eight knot (the rotations of the Planck time, space, charge, mass, and temperature boundaries), $G_{Gi} =$ Gieseking's constant, $n = 5$ the number of unique rotations partitioning the hyperbolic figure eight knot, $W_{We} =$ the Weierstrass constant, and $\boxplus$ = the terminal boundary of the hyperbolic figure eight knot.

The gamma function elegantly defines the partitioned balance of time, space, charge, mass, and temperature. The "most beautiful function of all" characterizes how the boundaries of constructive reality (the Planck boundaries) are maintained under ideal hyperbolic factorization. This single equation reveals the 5 boundaries of physical reality as the unique cyclic expressions of the minimal geometric form. It defines reality's algebraic structure in terms of the simplest 3-manifold, giving us rich intuitive access to the internal constructive balance of existence.

Before we explore how to use this new functional insight to discover reality's beautiful intricacies, let's examine the inversion synchronicity between the decomposed balance of the hyperbolic figure eight knot and the gamma function, by breaking the gamma function up into its parts, graphing those parts, and comparing them to our previous graphs of the internal actions of the hyperbolic figure eight knot.

First, we notice that under unitary and inverse-complex argument, the Q regularized gamma function $Q(1,(x+iy)^{-1})$ (Graph 12) is the inverse graph of the internal action of balance 1 (Graph 3), except for some small detail around (0,0).

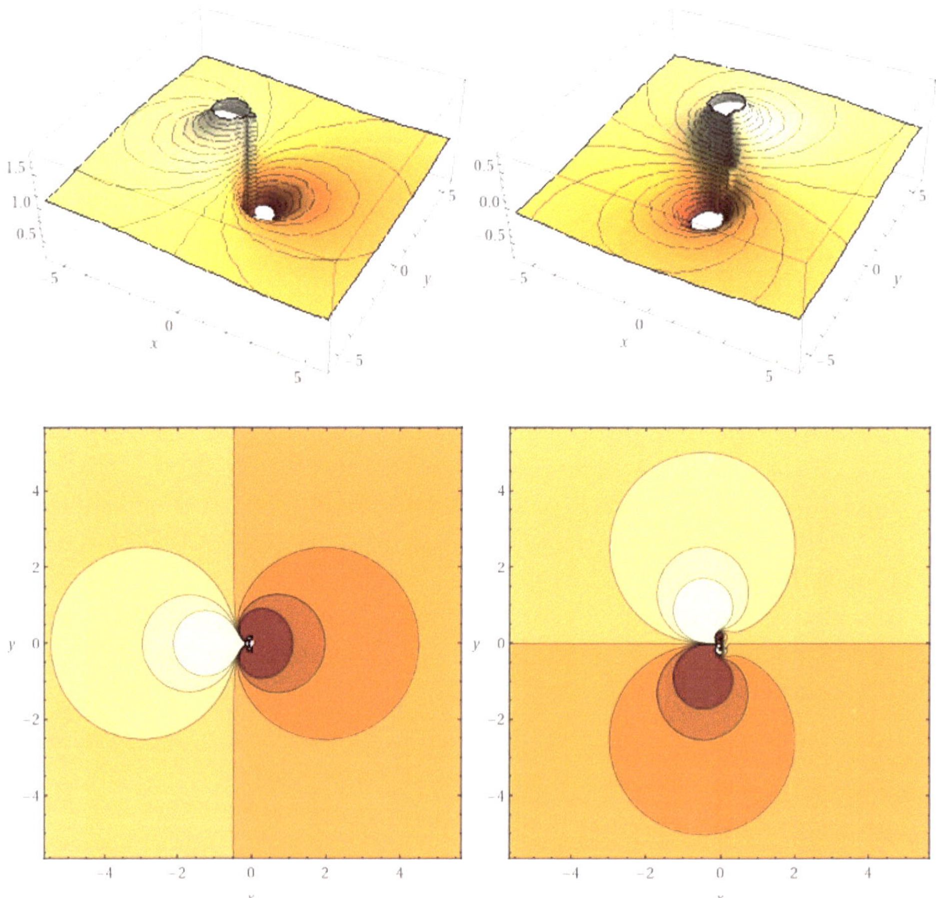

Graph 12: real (left) and imaginary (right) plots of the Q regularized gamma function under unitary and inverse-complex argument. $Q(1,(x+iy)^{-1})$. *Compare to balance 1 (Graph 3).*

Zooming in on the detail around (0,0) (Graph 13) we find a reproduction of the internal action of balance 2 (Graph 6), dubbed "the partition butterfly".

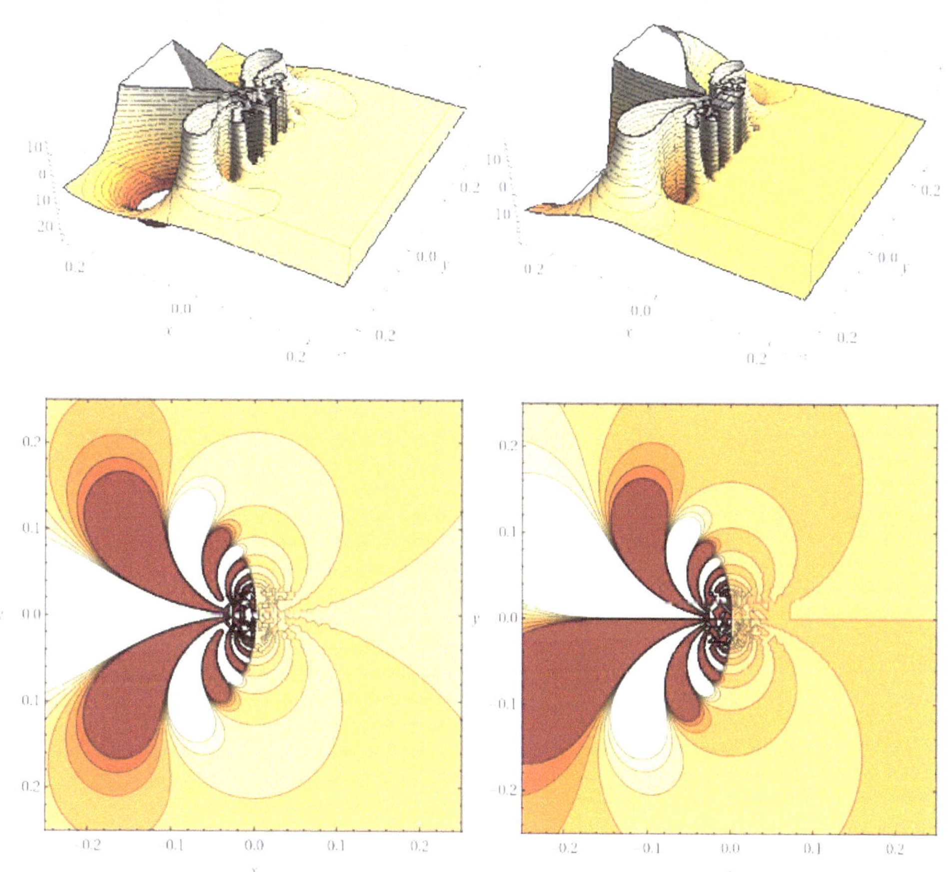

Graph 13: real (left) and imaginary (right) plots of the Q regularized gamma function under unitary and inverse-complex argument, zoomed in. $Q(1,(x+iy)^{-1})$ *Compare to balance 2 (Graph 6). "the partition butterfly"*

This unitary, inverse-complex connection of the Q regularized gamma function shows the trivial asymmetric connection between balance 1 and 2.

Zooming out, we see that the unitary and complex argument the Q regularized gamma function $Q\big(1,(x+iy)\big)$ (Graph 14) reproduces the internal action of balance 2 (Graph 5).

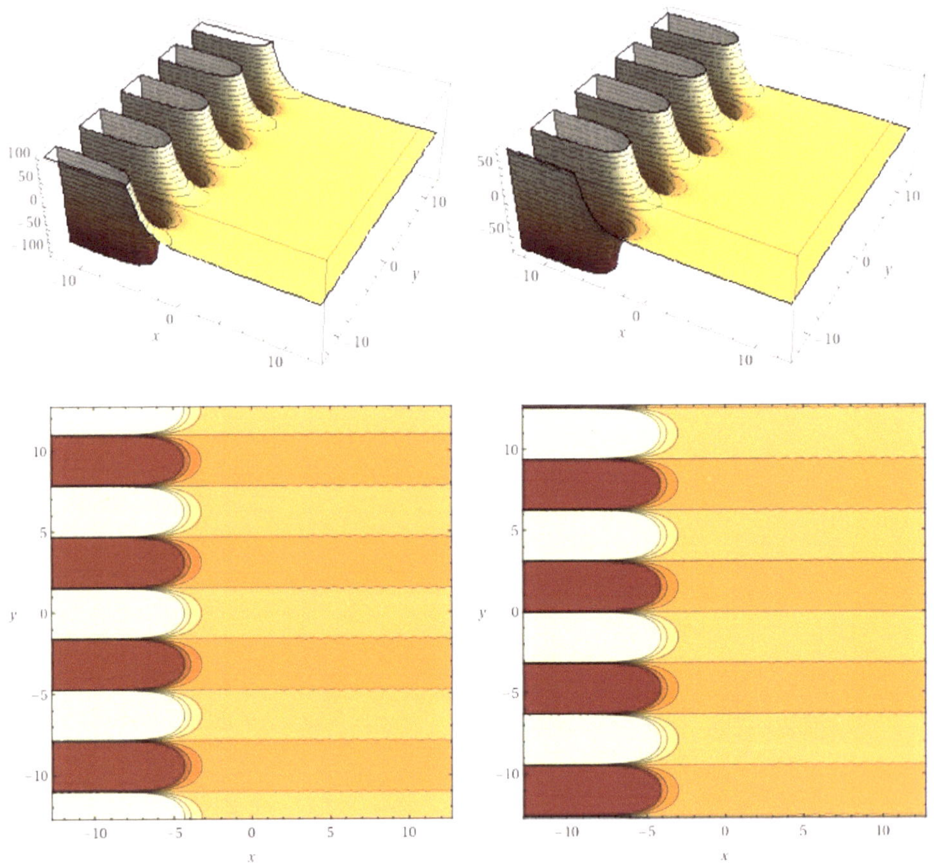

Graph 14: real (left) and imaginary (right) plots of the Q regularized gamma function under unitary and complex argument $Q\big(1,(x+iy)\big)$*. Compare to balance 2 (Graph 5).*

Under triple (unitary, complex, and inverse-complex) arguments the P regularized gamma function $P(1,(x+iy),(x+iy)^{-1})$ maps the self-dual external (real) and internal (imaginary) partition space. Note the unit circle in the real factorization map (Graph 15).

An awareness of the gamma function's central role in the persistent balance of reality, coupled with an awareness of the binomial construction of that balance, suddenly equips us with the ability to ask many new precise questions about reality and to discover their geometric solutions. For example, we can ask, "On what boundaries does the gamma function factor? And what is the geometry of those factorizations?"

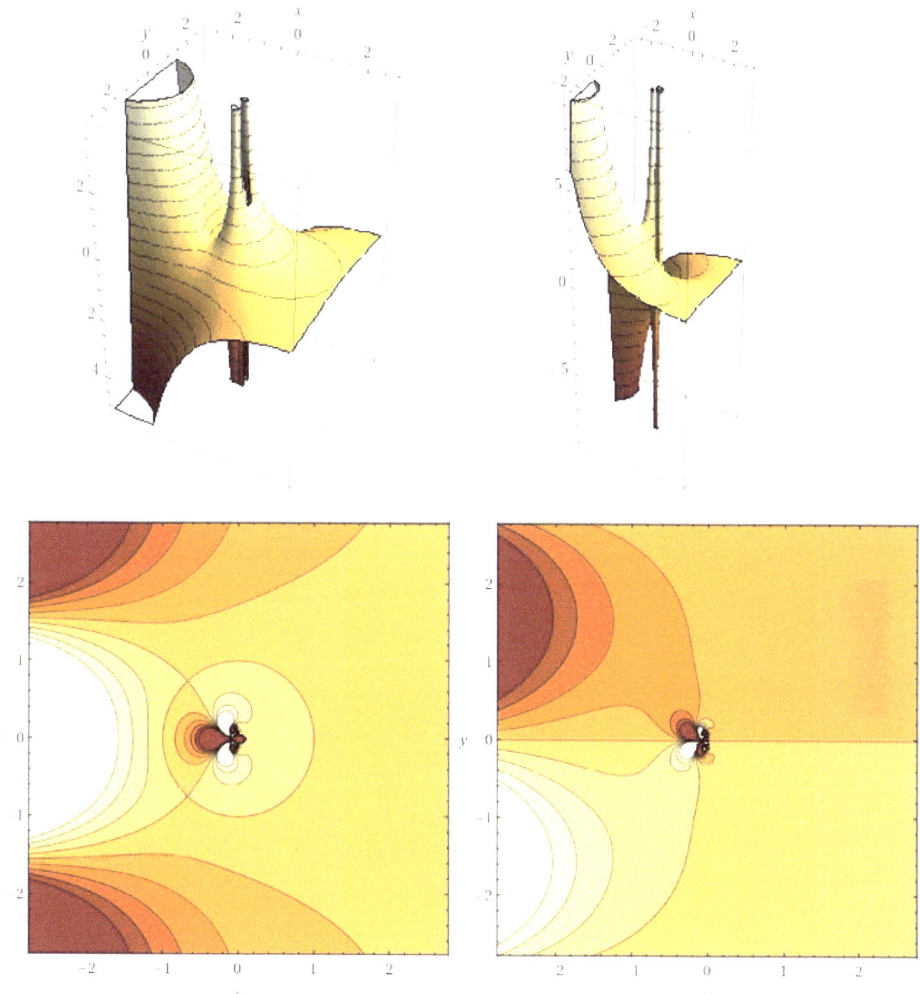

Graph 15: real (left) and imaginary (right) plots of the P regularized gamma function under unitary, complex, and inverse-complex argument $P(1,(x+iy),(x+iy)^{-1})$.

To find the gamma function's trivial factorization boundary we set the gamma function's argument to the sum of the hyperbolic figure eight knot's 5 rotations. Then we ask, "On what boundary B_p and with what terminal geometric action A_t does this set of factorizations trivially close (become equal to its own complete functional inverse)?"

$$\Gamma\left(\sum_{k=0}^{4} \phi_k\right) B_p = \int_0^{\infty} \frac{1}{\Gamma(x)}\, dx\, (1 + A_t \boxplus)$$

Where the sum of rotations is,

$$\sum_{k=0}^{4} \phi_k = \phi_0 + \phi_1 + \phi_2 + \phi_3 + \phi_4 = 12.4762773413029 \ldots$$

and the gamma function's complete inverse is defined as the integral of the inverse gamma function from zero to infinity, known as the Fransén-Robinson constant (F_{FR}).

$$F_{FR} = \int_0^\infty \frac{1}{\Gamma(x)} dx$$

Just like before, this equation has 2 unknowns, and since the terminal boundary of this geometry doesn't come into play until the 7th digit of this action, those 2 unknowns can be solved in pieces. That is, since ⊞ $= \left(\frac{l_P\, m_P}{q_P{}^2}\right) = 1 \times 10^{-7}$ we can safely ignore the geometric contribution of the terminal factor, solve for its primary boundary B_p, and then solve for the terminal geometric action A_t of that factorized balance.

By temporarily setting $A_t = 0$ and solving for B_p (expecting an answer that is correct to approximately the first 7 digits) we get:

$$\Gamma\left(\sum_{k=0}^{4} \phi_k\right) B_p = \int_0^\infty \frac{1}{\Gamma(x)} dx \qquad B_p = 2.17636464329325 \ldots \times 10^{-8}$$

$$m_P = 2.17642683817579 \ldots \times 10^{-8}$$

Which is a great approximation of the Planck mass boundary m_P.

Setting $B_p = m_P$, we solve for the terminal geometric action A_t of this trivial partition balance, finding that it divides the full derangements of this geometry ($!\boldsymbol{n}$) by the number of its unique partition rotations ($\boldsymbol{n}$) and maintains that division under square/square-root hyperbolic vortex lemniscate balance ($ж_r{}^2\sqrt{L}$).

$$A_t = ж_r{}^2\sqrt{L}\left(\frac{!\boldsymbol{n}}{\boldsymbol{n}}\right)$$

From this we conclude that the gamma function's primitive factorization boundary is the Planck mass boundary, and that this boundary is terminally maintained under hyperbolic-lemniscatic factorization balance.

$$\Gamma\left(\sum_{k=0}^{4} \phi_k\right) m_P = \int_0^{\infty} \frac{1}{\Gamma(x)} dx \left(1 + ж_r{}^2 \sqrt{L} \left(\frac{!n}{n}\right) ⊞\right)$$

Where $\Gamma(x)$ = the gamma function, m_P = the Planck mass, $ж_r$ = the hyperbolic vortex radius constant, and L = the lemniscate constant.

To ask about the geometry of the gamma function's other 2 external factorization boundaries (q_P and T_P) we follow the same procedure for the *squared* and *fourth* powered gamma factorizations of the same rotations. That is, we set the boundaries of the squared and fourth power gamma factorizations equal to the Planck charge and the inverse Planck temperature boundaries. Then for each we ask, "What is the primary action of this factorization, and what is its terminal action?"

$$\Gamma\left(\sum_{k=0}^{4} \phi_k\right) m_P = \int_0^{\infty} \frac{1}{\Gamma(x)} dx \left(1 + ж_r{}^2 \sqrt{L} \left(\frac{!n}{n}\right) ⊞\right)$$

$$\Gamma^2\left(\sum_{k=0}^{4} \phi_k\right) q_P = A_{p_2} \left(1 + A_{t_2} ⊞\right)$$

$$\Gamma^4\left(\sum_{k=0}^{4} \phi_k\right) \frac{1}{T_P} = A_{p_4} \left(1 + A_{t_4} ⊞\right)$$

Where $\Gamma(x)$ = the gamma function, $\sum \phi_k$ = the sum of unique rotations composing the hyperbolic figure eight knot, m_P, q_P, and T_P = the Planck mass, charge, and temperature boundaries, and ⊞ = the terminal boundary of the hyperbolic figure eight knot.

Solving first for the primary and then the terminal action of each, we conclude that: the sum of the unique rotations composing the hyperbolic figure eight knot trivially factor about the Planck mass boundary, the square of those rotations elliptically factor about the Planck charge boundary, and the 4th power of those rotations hyperbolically factor about the inverse Planck temperature boundary.

factorization order × boundary = binomial factorization geometry

$$\Gamma\left(\sum_{k=0}^{4} \phi_k\right) m_P = \int_0^{\infty} \frac{1}{\Gamma(x)}\, dx \left(1 + ж_r{}^2 \sqrt{L} \left(\frac{!n}{n}\right) ⊞\right) \qquad \text{primitive}$$

$$\Gamma^2\left(\sum_{k=0}^{4} \phi_k\right) q_P = \left(\frac{2}{3}\right) n\, \sigma_{zs}{}^4 \left(1 + \left(\frac{ж_2}{ж_1}\right)^2 \left(\frac{2}{3}\right) \left(\frac{\log 2}{\log 3}\right)^{3/2} ⊞\right) \qquad \text{elliptic}$$

$$\Gamma^4\left(\sum_{k=0}^{4} \phi_k\right) \frac{1}{T_P} = (n - 3G_{Gi}) \left(1 + \frac{n}{3} \left(2b\, C_{CFP}{}^2\right)^2 ⊞\right) \qquad \text{hyperbolic}$$

Where $\Gamma(x)$ = the gamma function, $\sum \phi_k = \phi_0 + \phi_1 + \phi_2 + \phi_3 + \phi_4$, m_P, q_P, and T_P = the Planck mass, charge, and temperature boundaries, $n = 5$ the number of unique rotations partitioning the hyperbolic figure eight knot, $!n$ = the number of derangements created by those rotations, $ж_1$ and $ж_2$ are the 1st and 2nd hyperbolic vortex partition constants, $ж_r = \sqrt{ж_3 ж_4}$ = the hyperbolic vortex radius constant, L = the lemniscate constant, G_{Gi} = Gieseking's constant, C_{CFP} = the real fixed point of the hyperbolic cotangent, ⊞ = the termination boundary of the hyperbolic figure eight knot, and σ_{zs} = the Zolotarev-Schur constant.

$$\sigma_{zs} = \frac{1}{c^2}\left[1 - \frac{E(c)}{K(c)}\right]^2$$

Where $0 < c < 1$ = the unique solution of

$$[K(c) - E(c)]^3 + (1 - c^2)K(c) - (1 + c^2)E(c) = 0$$

$K(c)$ = the complete elliptic integral of the 1st kind, and $E(c)$ = the complete elliptic integral of the 2nd kind.

The universal binomial factorization prescription is a powerful new tool for waking ourselves up to the always present features of existence we've remained blind to. Under that prescription, every clear question we ask the gamma function makes us aware of a unique geometric feature of persistence, enabling us to see features of reality that have always been there, but remained hidden from our conscious view (like seeing the number of an object's sides was hidden from view before Euler).

For example, if we ask, "How does the time boundary factor ($\Gamma(\phi_0)$)?" the universal binomial factorization prescription tells us that it factors into seven circles ($2\pi\ \boldsymbol{b}$) terminally arranged into a split squared balanced derangement ($!\boldsymbol{n}$) of the minimum manifold.

$$\Gamma(\phi_0) = 2\pi\ \boldsymbol{b}\left(1 + \left(\frac{\boldsymbol{n}^2}{\sqrt{!\boldsymbol{n}}}\right)\left(\frac{ж_r{}^2}{G_{Gi}}\right)^2 \boldsymbol{b}^{-1}\ \boxplus\right)$$

Where $\Gamma(x)$ = the gamma function, $ж_r$ = the hyperbolic vortex radius constant, G_{Gi} = Gieseking's constant, $\boldsymbol{n} = 5$ the number of unique rotations partitioning the hyperbolic figure eight knot, $\boldsymbol{b} = 7$ the break in scale symmetry of the hyperbolic figure eight knot's internal complement, and $\boxplus$ = the termination boundary of the hyperbolic figure eight knot.

This enriched understanding of the gamma function carries over into the function that sits at the heart of the Riemann hypothesis, the Riemann zeta function.

- -

The reader is strongly encouraged to continue exploring reality's constructive secrets by asking precise questions of the gamma function and using the universal binomial prescription to find their answers. For example, you might ask:

question 1

$$\Gamma\left(\frac{\phi_4}{\phi_2\ \phi_3}\right) = \left(\frac{\sqrt{\boldsymbol{b}}}{G_{Gi}{}^2}\right)\left(1 - ж_2\left(cos\left(sinh\left(\frac{4}{\boldsymbol{b}}\right)\right)^{-1}\right)^{-1}\ \boxplus\right)$$

Where $\Gamma(x)$ = the gamma function, ϕ_2, ϕ_3, and ϕ_4 = the rotations of the Planck charge mass and temperature boundaries, G_{Gi} = Gieseking's constant, $cos(x)$ = the cosine function, $sinh(x)$ = the cosine function, $ж_2$ = the 1st hyperbolic vortex partition constant, and $\boxplus$ = the termination boundary of the hyperbolic figure eight knot.

question 2

$$\Gamma(0, \phi_2) = \left(\frac{b}{2}\right)\frac{Im(\omega_1)^3}{\Gamma(n)}\left(1 + ж_r{}^2\, sinh\left(cosh\left(\frac{b}{4}\right)\right)\boxplus\right)$$

Where $\Gamma(0, x)$ = the incomplete gamma function, ϕ_2 = the rotation of the Planck charge boundary, the $ж_r$ = the hyperbolic vortex radius constant, $cosh(x)$ = the hyperbolic cosine function, $sinh(x)$ = the cosine function, ω_1 = the omega_1 constant, and $\boxplus$ = the termination boundary of the hyperbolic figure eight knot.

question 3

$$\Gamma\left(\frac{1}{\phi_0 + \phi_1 + \phi_2}\right) = 2b\left(Im\left(i^{i^{i^{\cdots}}}\right)\right)^{\frac{1}{2}}\left(1 + \left(\frac{ж_2}{ж_1}\right)D_{Do}{}^{-\frac{1}{4}}\boxplus\right)$$

Where $\Gamma(x)$ = the gamma function, ϕ_0, ϕ_1, and ϕ_2 = the rotations of the Planck time, length and charge boundaries, $i^{i^{i^{\cdots}}}$ = the infinite power tower of i—the imaginary root, $ж_1$ and $ж_2$ are the 1st and 2nd hyperbolic vortex partition constants, D_{Do} = the unique real root of the cosine function (the Dottie number), and $\boxplus$ = the termination boundary of the hyperbolic figure eight knot.

question 4

$$\Gamma\left(\frac{1}{\phi_3}\right) = T_{tet}\left(1 - \frac{b^3}{(3Re(\omega_1)^n)^2}\boxplus\right)$$

Where $\Gamma(x)$ = the gamma function, ϕ_3 = the rotation of the Planck mass boundary, T_{tet} = the Tetranacci constant (the second polynomial root of the Tetranacci numbers), ω_1 = the omega_1 constant, and $\boxplus$ = the termination boundary of the hyperbolic figure eight knot.

$$T_{tet} = (x^4 - x^3 - x^2 - x - 1)_2 \qquad T_{tet} + T_{tet}{}^{-4} = 2$$

Chapter 13: the Riemann zeta function

Since the time of Hilbert, the dream has been to find a natural self-adjoint operator whose spectrum yields the zeros of the Riemann zeta function—whose zeros intrinsically relate to the distribution of prime numbers.[7]

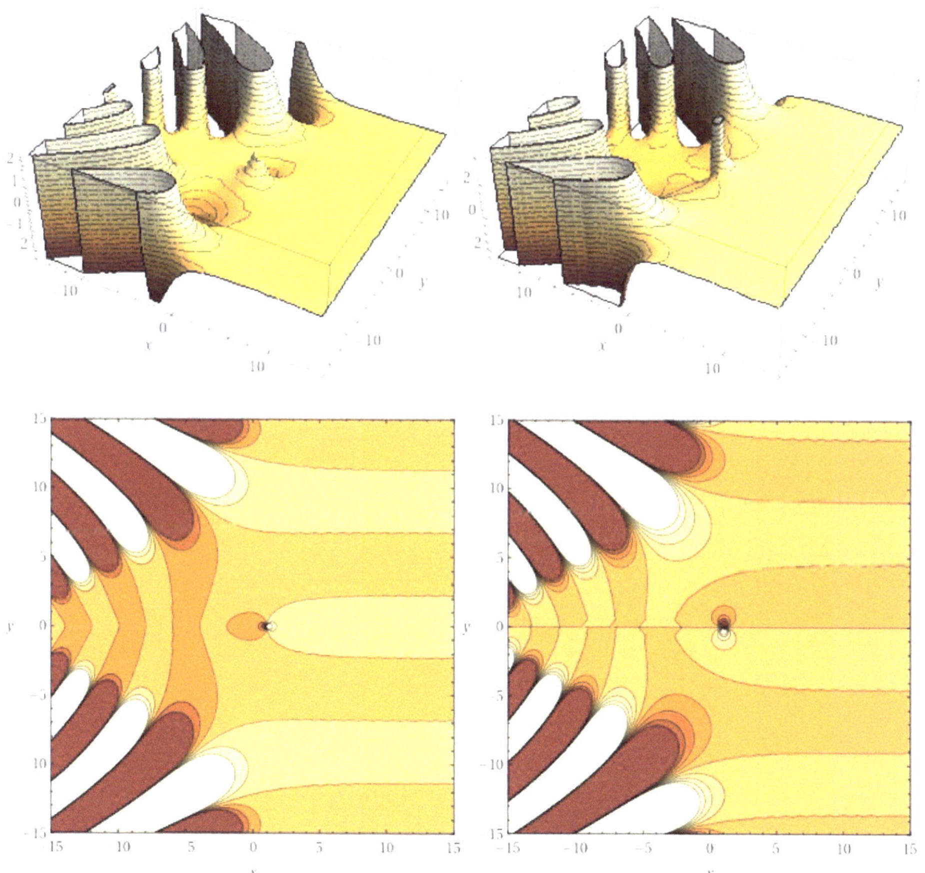

Graph 16: real (left) and imaginary (right) plots of the Riemann zeta function under complex argument $\zeta(x + iy)$.

[7] The Riemann hypothesis, widely regarded the most important unsolved problem in pure mathematics, conjectures that the zeros of the Riemann zeta function occur only at negative even integers, and at positive complex values $s = (x + iy)$ with real part $x = \frac{1}{2}$.

The Riemann zeta function is defined as an infinite *sum* and as an infinite *product* over the primes.

$$\zeta(s)=\sum_{n=1}^{\infty}\frac{1}{n^s} \qquad \zeta(s)=\prod_{p}\frac{1}{1-p^{-s}}$$

Where $s=(x+iy)$ is a complex number, and p is a prime number.

When $Re(s)=x>1$ the Riemann zeta function can be represented in terms of the inverse gamma function multiplied by the converging summation (integral) form of the gamma function offset by 1.

$$\zeta(s)=\frac{1}{\Gamma(s)}\int_0^{\infty}\frac{x^{s-1}}{e^x-1}dx \qquad Re(s)=x>1$$

bounded by $y\geq 3$, and $x\geq 1-\dfrac{1}{(log|y|)^{\frac{2}{3}}(loglog|y|)^{\frac{1}{3}}\left(1+4\,Im(\rho_1)\right)}$

Where $s=(x+iy)$, $Im(\rho_1)=$ the imaginary part of the first nontrivial root of the zeta function ρ_1, and $\Gamma(\mathrm{s})=$ the gamma function, which takes the integral form.

$$\Gamma(\mathrm{s})=\int_0^{\infty}\frac{x^{s-1}}{e^x-0}\,dx \qquad Re(s)>0$$

$\rho_1=0.5+14.1314251417346\ldots i$ 1st nontrivial zero of the zeta function

Special known values of the zeta function include:

$$\zeta(0)=-\frac{1}{2} \qquad \zeta'(0)=-\frac{1}{2}\log(2\pi) \qquad \zeta(1)=\infty$$

$$\zeta(2)=\frac{\pi^2}{2\,(3)} \qquad \zeta(4)=\frac{\pi^4}{2\boldsymbol{n}\,(3)^2} \qquad \zeta(6)=\frac{\pi^6}{(2+\boldsymbol{n})\boldsymbol{n}\,(3)^3}$$

Graphing the Riemann zeta function under inverse complex argument we notice it contains split binding points, maintaining separate junctions about (0,0) and (1,0) (Graph 17).

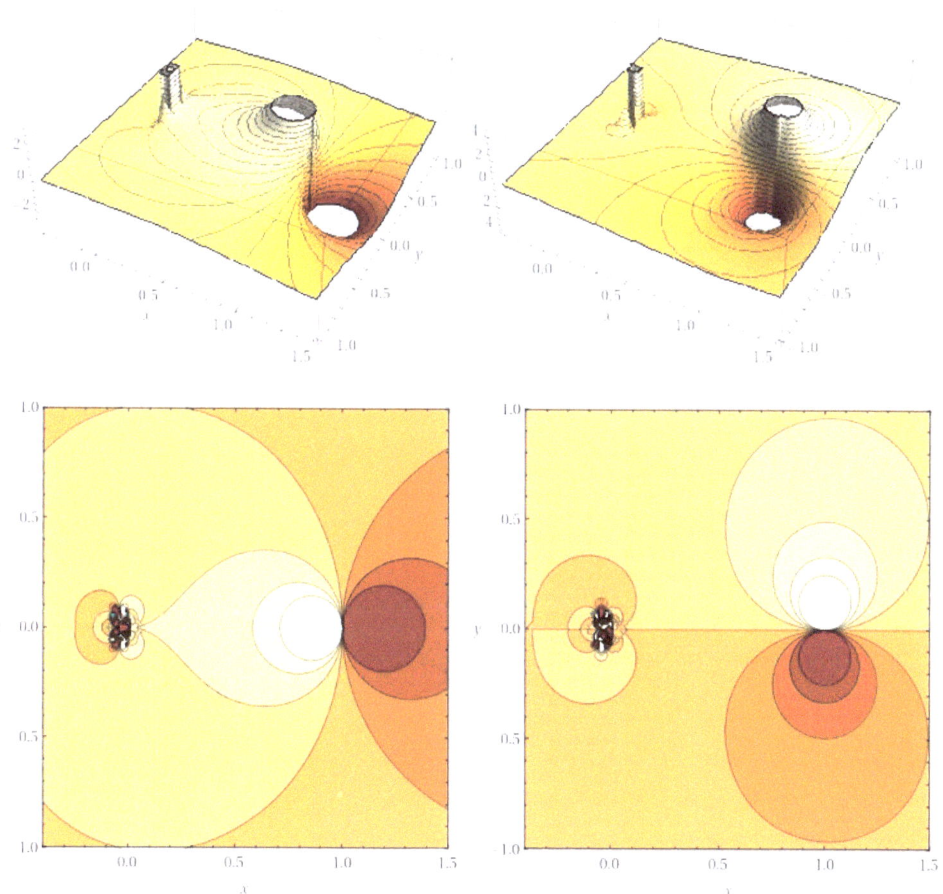

Graph 17: real (left) and imaginary (right) plots of the Riemann zeta function under inverse-complex argument $\zeta((x+iy)^{-1})$.

Zooming in on the detail around (0,0) (Graph 18) gives us a reproduction of the internal action of balance 2 (Graph 6). And zooming in on the detail around (1,0) (Graph 19) reproduces the inverse graph of the internal action of balance 1 (Graph 3).

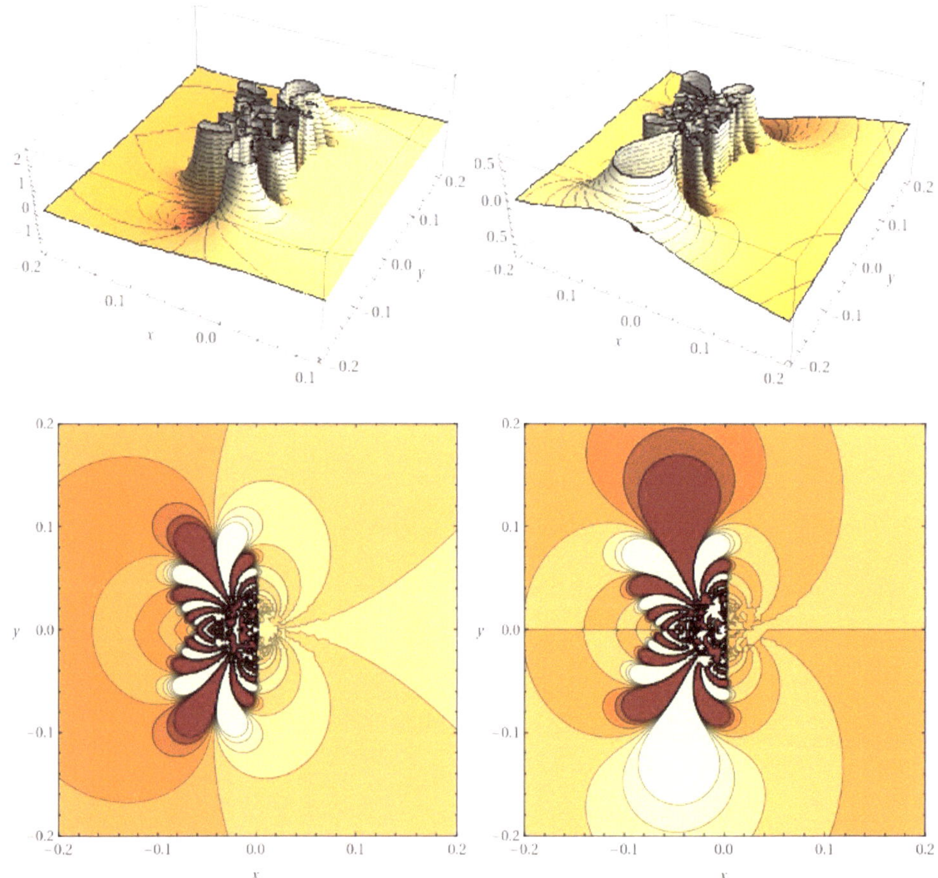

Graph 18: real (left) and imaginary (right) plots of the Riemann zeta function under inverse-complex argument $\zeta((x+iy)^{-1})$, *zoomed in around* $(0,0)$.

It has long been known that the key to solving the Reimann hypothesis and unraveling the deep mysteries of the primes lies in finding a self-adjoint operator that is naturally bound by the zeta function's zeros. That is, finding a geometric explanation for the Reimann zeta function. Since the trivial zeros are not difficult to see the pattern in, the dream has been to find the geometric reason for why the non-trivial zeros (which all have real part ½) have the values they do. Most importantly, why does the first non-trivial zero of the zeta function $Im(\rho_1)$ has the value it has?

Using the universal binomial factorization prescription, $Im(\rho_1)$ is trivially revealed as the circular/hyperbolic factorization boundary that terminally maintains the hyperbolic vortex radius ($ж_r$) under simple chiral connection.

$$ж_r = Im(\rho_1)\left(\cos\left(\cosh\left(\frac{n}{2}\right)\right)\right)^{-1}\left(1-\frac{n}{2}\left(3^2\, Re\left(i^{i^{i^{\cdots}}}\right)\right)⊞\right)$$

Where $ж_r$ = the hyperbolic vortex radius constant, $Im(\rho_1)$ = the imaginary part of the first non-trivial zero of the zeta function, $cos(x)$ = the cosine function, $cosh(x)$ = the hyperbolic cosine function, $n = 5$ the number of unique rotations partitioning the hyperbolic figure eight knot, $Re\left(i^{i^{i^{\cdots}}}\right)$ = the real part of the infinite power tower of i—the imaginary root, and ⊞ = the terminal boundary of the hyperbolic figure eight knot.

In other words, the first non-trivial zero of the Riemann zeta function constructively defines the zero bound (throat) of the hyperbolic vortex.

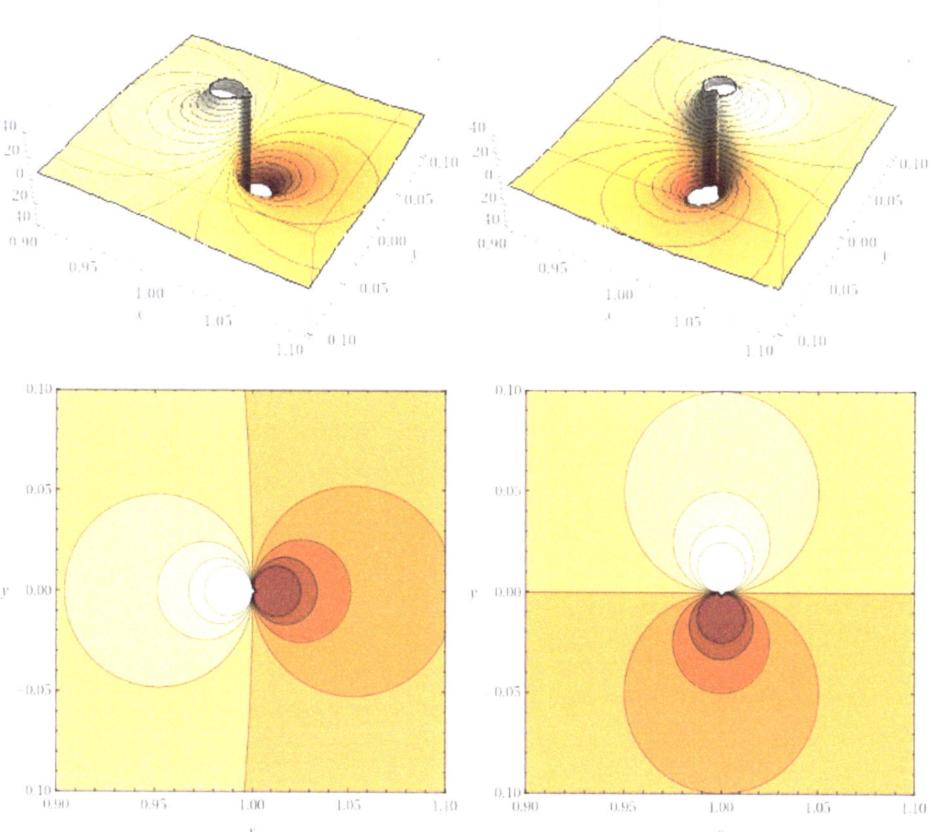

Graph 19: real (left) and imaginary (right) plots of the Riemann zeta function under inverse-complex argument $\zeta((x+iy)^{-1})$, *zoomed in around* $(1,0)$. *Compare to Graph 3.*

The gamma function and the zeta function are ideal tools for exploring the construction of reality via the universal binomial prescription. Every question these functions answer sharpens our geometric understanding of the partition balance of reality, defining a unique feature of persistence.

- -

The reader is strongly encouraged to explore the constructive secrets of reality by asking precise questions of the Reimann zeta function and using the universal binomial prescription to find their answers.

For example, for a richer understanding of the zeta function we might ask:

Question 1: What is the zeta function value of the time boundary's rotation?

$$\zeta(\phi_0) = 3\,Re(\omega_1)^4 \left(1 + 3\,\Gamma(\boldsymbol{n})\left(\,\omega_2\,\Gamma(x_{min})\right)^4 ⊞\right)$$

Where $\zeta(x)$ = the zeta function, ϕ_0 = the rotation of the time boundary, ω_1 and ω_2 = the omega_1 and omega_2 constants, $\Gamma(x)$ = the gamma function, $\Gamma(x_{min})$ = the minimum value of the gamma function for positive argument, and ⊞ = the terminal boundary of the hyperbolic figure eight knot.

Question 2: What is the zeta function value of the space boundary's rotation?

$$\zeta(\phi_1) = 3\,m_R\left(1 + ж_r{}^2\left(\frac{\Gamma(\boldsymbol{n})}{!\boldsymbol{n}\,W(1)^4}\right)^3 ⊞\right)$$

Where $\zeta(x)$ = the zeta function, ϕ_1 = the external rotation of the space boundary, $ж_r$ = the hyperbolic vortex radius constant, $\Gamma(x)$ = the gamma function, ⊞ = the terminal boundary of the hyperbolic figure eight knot, m_R = Rényi's parking constant,

$$m_R = e^{-2\gamma}\int_0^\infty \frac{e^{-2\Gamma(0,x)}}{x^2} = e^{-2\gamma}\int_0^\infty \frac{e^{2Ei(-x)}}{x^2}$$

And the omega constant $W(1)$ = is the unitary value of Lambert's W function, defined as the unique real number that satisfies the equation $\Omega e^{\Omega} = 1$.

Question 3: What is the zeta function value of the Planck length's inverse square rotation?

$$\zeta\left(\phi_1{}^{-2}\right) = -\left(\frac{2}{3}\, e^{\gamma}\right)^{\frac{1}{2}} \left(1 + 2n\,\left(3V_{fe}\right)^2 ⊞\right)$$

Where $\zeta(x)$ = the zeta function, ϕ_1 = the external rotation of the space boundary, γ = the Euler-Mascheroni constant, V_{fe} = the volume of the hyperbolic figure eight knot complement, and ⊞ = the terminal boundary of the hyperbolic figure eight knot.

Question 4: What is the zeta value of inverse hyperbolic vortex radius division?

$$\zeta\left(\frac{1}{ж_r}\right) = -4\pi\, sinh^{-1}\left(\frac{1}{4}\right)^2 \left(1 + \left(2\,ж_2\, G_g\right)^2 ⊞\right)$$

Where $ж_1{}^2 + ж_2{}^2 + ж_3{}^2 + ж_4{}^2 = -4\pi$, and G_g = the tether length at which a goat tied to the boundary of a unit circular field can graze exactly half the field.

Question 5: What is the zeta value of square inverse hyperbolic vortex radius division?

$$\zeta\left(\frac{1}{ж_r{}^2}\right) = -\frac{bn}{C_{Murata}{}^4}\left(1 + \frac{\pi}{8} cosh^2(4)\, ⊞\right)$$

Where C_{Murata} = the Murata constant defines the following product over the primes.

$$C_{Murata} = \prod_p \left(1 + \frac{1}{(p-1)^2}\right)$$

Where p = a prime number.

Chapter 14: inside

There are exactly 4 kinds of numbers that can be added, subtracted, multiplied and divided. That is, there are exactly 4 division algebras: real numbers, complex numbers, quaternions, and octonions. The real numbers are 1-dimesnional. Pairs of real numbers make the complex numbers, which are 2-dimensional. Likewise, pairs of complex numbers make the quaternions (4-D), and pairs of quaternions make the octonions (8-D). And that's the end of the line. That is, the octonions represent the last of the division algebras. No higher order of pairings will allow all those symmetric operations.

real ($\mathbb{R}$)	1D
complex ($\mathbb{C}$)	2D
quaternion ($\mathbb{H}$)	4D
octonion ($\mathbb{O}$)	8D

These division algebras reflect the constructive division balances assigned to the different domains of reality. The external domain connects under real, complex, and quaternion construction, whereas the internal domain rearranges those same parts into octonion construction.

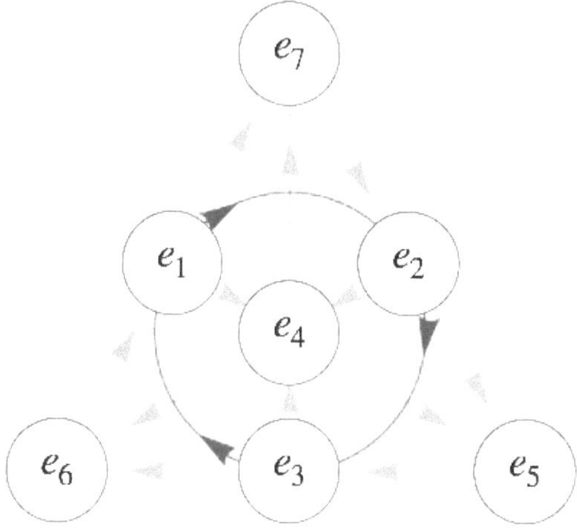

Cayley logic table for the octonions portraying the 7 basis vectors e_k and their algebraic connections.

In total, the octonion logic symmetrically connects 8 things (7 basis vectors plus the unitary element 1) under closed 3-member operations. This means it topologically decomposes into 8 simply connected triangles, or 2 tetrahedra (4 simply connected triangles make a tetrahedra).

To visualize what the full octonion connection looks like (its 8 symmetric triple groupings) we graph the product of the inner actions of balance 0 and balance 1.

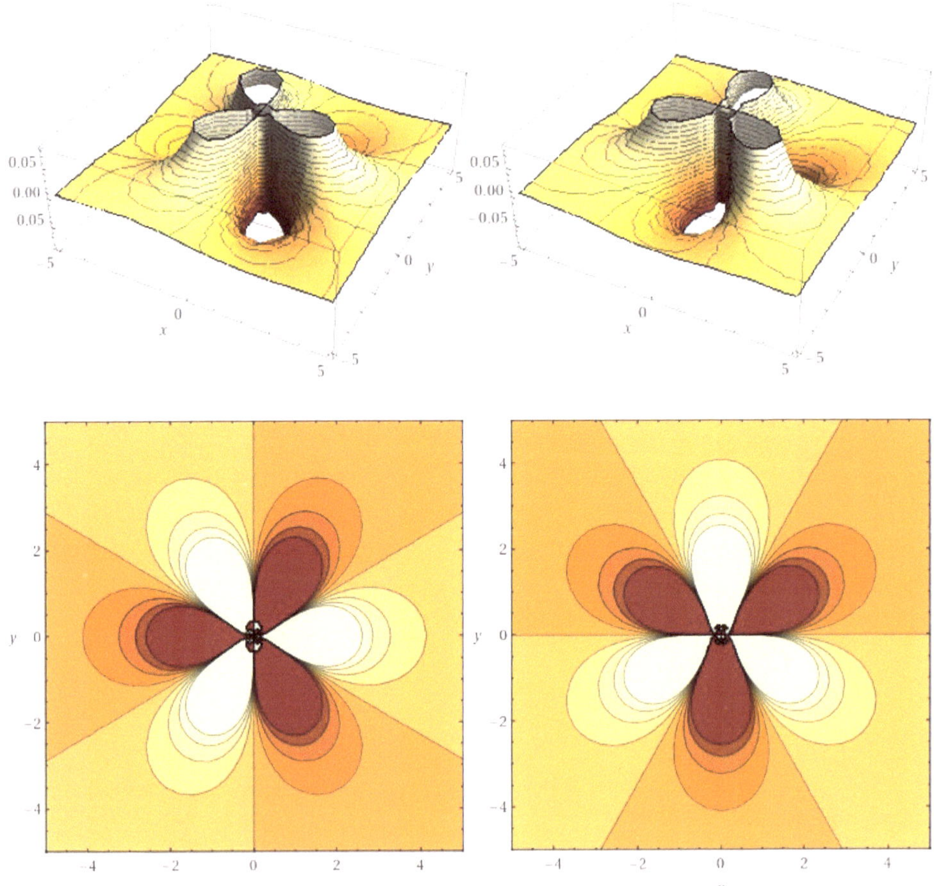

Graph 20: real (left) and imaginary (right) plots of the product of the internal actions of balance 0 and 1, each under inverse-complex argument. Compare to the Cayley logic table for the octonions.

The *real* and *complex* graphs both identify 2 inner groups of 3 and 2 exterior groups of 3 connecting under 5-part asymmetric balance; the 8 3-member simply connected operations of the closed the octonion logic.

Chapter 15: outside

The minimum geometric balance has now been defined in terms of minimal derangements and in terms of hyperbolically balanced factorizations (the gamma function). In the last chapter we noted that the *internal* geometry of that minimum volume partitioning is bound under octonion logic. In this chapter we examine the *external* geometric expression of this balance (defining the n-hypersphere of maximal volume) and note its connective logic.

To an *internal* observer the universe appears bounded into the shape of the hyperbolic figure eight knot. *Externally*, the constructive parts of those boundaries are rearranged into the shape of the n-hypersphere of maximal volume. What does the n-hypersphere of maximal volume look like?

On its grandest scales, the n-hypersphere of maximal volume defines the simplest possible geometric form in $\mathbb{S}^3$—*the great sphere* minus its boundary. On the other end of things, the internal boundary of the n-hypersphere of maximal volume defines the second simplest geometric form, the minimal hypersurface known as the Clifford torus.

As a helicoidal product surface, the Clifford torus defines the simplest and most symmetric flat embedding of the cartesian product of two circles. That is, it defines the simplest connection between 2 circles, each of which possess their own independent embedding space $\mathbb{R}^2_a$ and $\mathbb{R}^2_b$, resulting in a product space that is $\mathbb{R}^4$ (the 4-dimensional hyperbolic spacetime of general relativity).[8]

But that's not the end of the story. Since this helicoidal surface connects 2 circles of significantly different size, the chiral connection of its rotation operator is highly asymmetric, causing the cartesian product of these two circles to twist into an $\mathbb{R}^3$ torus.

Under asymmetric chiral connection this is unavoidable. The first circle in that connection consumes x and y leaving only one independent axis z available to the second circle. This quickly collapses the $\mathbb{R}^4$ domain into a $\mathbb{R}^3$ projection. "A torus embedded in $\mathbb{R}^3$ is an asymmetric reduced-dimension projection of the maximally symmetric Clifford torus embedded in $\mathbb{R}^4$."[9]

---

[8] Since $\mathbb{R}^4$ is identified with $\mathbb{C}^2$, this is equivalent to saying that the 3-sphere $\mathbb{S}^3$ lives in $\mathbb{C}^2$.

[9] https://en.wikipedia.org/wiki/Clifford_torus

The equation for the tangent cone of the Clifford torus is

$$x_1x_2 = x_3x_4$$

Generalized equation for the tangent cone of the Clifford torus.

The division parameters the Clifford torus' tangent cone are the connectors of the hyperbolic figure eight knot's terminal boundary.

$$\phi_1\phi_3 = \phi_2\phi_2$$

Where ϕ_1, ϕ_2, and ϕ_3 = the rotations of the Planck length, charge, and mass boundaries. Note that, although this construction is 4-dimensional, only 3 of those dimensions are unique.

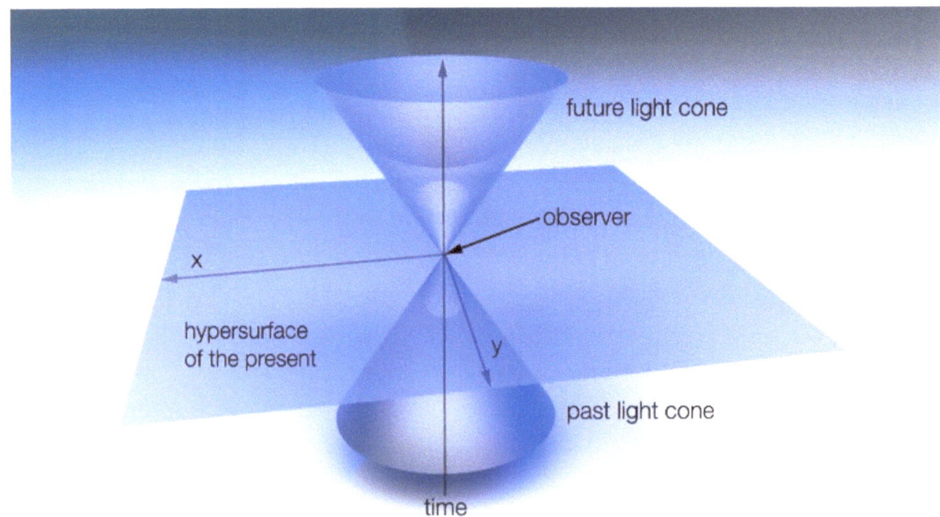

The tangent cone of reality's external Clifford torus projection.

The tangent cone of the Clifford torus defines the projective *null cone* of hyperbolic geometry. The entire surface of this hypercone defines the division singularities of the external domain. And the balance maintained by this constructive arrangement (between the future hypercone and the past hypercone) defines the *hyperplane* of the present—the most trivial minimal hypersurface.

The other minimal hypersurfaces embedded in this geometry (those that play a role in the helicoidal collapse from $\mathbb{R}^4$ to $\mathbb{R}^3$) are found by generalizing the Clifford torus via the method of separation of variables.[10]

minimal hypersurfaces

Clifford torus
tangent cone
hyperplane
lemniscate
minimum catenary

> "These are all the minimal hypersurfaces that can be obtained by the method of separation of variables."
>
> Jaigyoung Choe[11]

Since the external domain is maintained under n-hypercube connection (minimally bounded by the Clifford torus), it comes trivially programmed with the rules of differentiation and integration.[12] The rules of calculus define how the external part of this balance connects.

By definition, the *derivative* defines how the volume of a n-dimensional hypercube increases or decreases as the side length of that hypercube is changed. On the other hand, "integrating this picture—stacking the faces—geometrizes the fundamental theorem of calculus,

---

[10] See Jaigyoung Choe. Some minimal submanifolds, generalizing the Clifford torus.

[11] Ibid.

[12] The Clifford torus is a Möbius torus. Going around the boundary once puts you on its other side of its surface, requiring 2 loops to return to the starting condition (see Heegaard splitting—embedded minimal surfaces in compact manifolds of positive sectional curvature). It is also an example of a square torus, because it is isometric to a square with opposite sides identified (making it a manifold). The square torus can also be embedded into three-dimensional space by the Nash embedding theorem. One possible embedding modifies the standard torus by a fractal set of ripples running in two perpendicular directions along the surface.

yielding a decomposition of the n-cube into n pyramids, which is a geometric proof of Cavalieri's quadrature formula."[13]

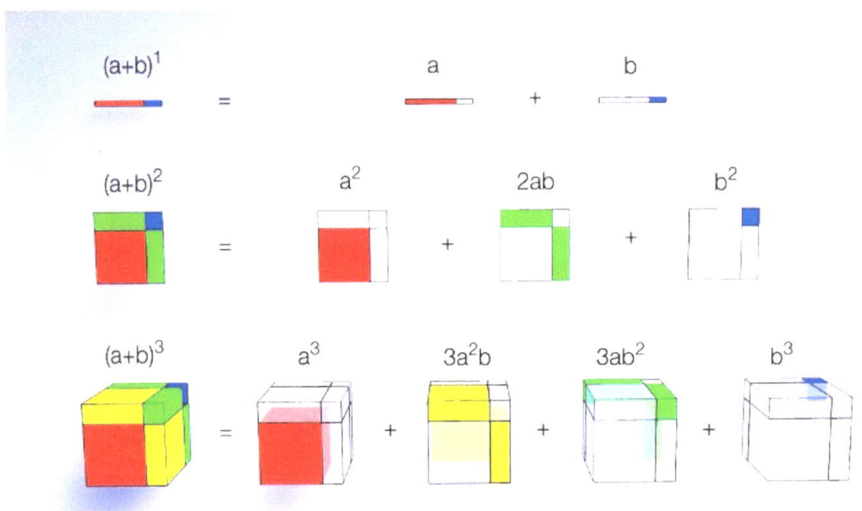

A geometric proof of the derivative and integral. A visualization of the binomial expansion up to the 3rd power.

The structure of calculus is also centrally scripted by the ideal hyperbolic connection (e) whose generalized power series defines the elementary *derivative sequence.*[14]

$$e^x = \frac{x^0}{0!} + \frac{x^1}{1!} + \frac{x^2}{2!} + \frac{x^3}{3!} + \frac{x^4}{4!} \cdots$$

The hyperbolic power series and the derivative. Every term in the generalized exponential series defines the derivative of the following term.

---

[13] Cavalieri's quadrature formula Wikipedia page. Note: the area, or "quadrature", of a hyperbola is a logarithm.

[14] The exponential function is also the unique nontrivial function that is its own derivative (up to multiplication by a constant) and its own antiderivative. In the graph $y = e^x$ the slope of the curve at any point x is e^x, and the area under the curve from negative infinity up to x is also e^x.

$$\frac{d}{dx} e^x = e^x \qquad \int e^x \, dx = e^x + C$$

The simplest possible self-balance naturally defines the division parameters of physics, and the arithmetic of this balance hierarchically maintains *all* the division algebras under closed asymmetric connection.

The internal geometry of this balance defines the minimum possible volume partitioning, the hyperbolic figure eight knot. Whereas its external projection defines the n-hypersphere of maximal volume.

On its largest scales the n-hypersphere of maximal volume defines the simplest possible geometric form in $\mathbb{S}^3$ the great sphere, minus its boundary. And on its smallest scales the n-hypersphere of maximal volume is internally bounded by the second simplest geometric form (the Clifford torus), which collapses from the $\mathbb{R}^4$ domain into the familiar $\mathbb{R}^3$ projection of common experience under asymmetric chiral connection. The connective logic of this external projection (the n-hypersphere of maximal volume) defines the rules of calculus.

Chapter 16: hyperbolized

In this book we have found that the minimal partitioned volume (the hyperbolic figure eight knot complement volume) and the n-hypersphere of maximal volume jointly form the minimum partitioned balance. This balance defines an arena maintained under the projection of 5 perpetual actions with Planck constant boundaries (time, space, charge, mass and temperature). The external *charge* and *mass* boundaries of that balance partition into the exact charge and mass values that define the fundamental particles of matter. And the boundary intersections of this minimal balance define the constants of Nature.

We have also found that this *theory of everything* constructively maintains a dual hyperbolic union, one that hyperbolically connects *and* hyperbolically partitions—defined by the gamma function and, orthogonally, by the Reimann zeta function. The external projection of this geometry defines the n-hypersphere of maximal volume, which takes the shape of the great sphere, minus its boundary, on its maximum scale, and on its smallest scales is internally bounded by the second simplest geometric form (the asymmetric Clifford torus). This n-hypersphere of maximal volume trivially maintains the connective logic of calculus.

Future editions of this book can be expected to add the following topics to this conversation.

Primes

Having a minimal derangement/factorization theory means having a formally constructive definition of the rules of the minimum stage, defining how the minimum projective system (the 1) fundamentally internally factors. This conceptual framework for what it means to be divisible by 1 (unit divisibility) puts us in the proper position for addressing the great mysteries of the primes, the set of unique division elements in reality.

Platonic solids—E8

This story should properly finish by wrapping back around to its beginning, to the Platonic solids and the idea that the universe was somehow composed of them. After all, Plato was *hyperbolically* right. Reality does have a shape that defines the parameters of its construction. That shape decomposes into 5 elementary geometric parts. Thurston showed that the inner complement of the hyperbolic figure eight knot decomposes into 2 regular hyperbolic tetrahedra. This almost begs us to complete the story, to tell the story for how the other elementary partitions relate to the icosahedron, and so on. A rich exploration of this topic from the perspective of exceptional symmetry groups (E7, E8, etc.) would also be of value.

connection logic?

> *"If you have a closed manifold with a surface in it, then its self-intersection number is the electric charge."*
>
> *Sir Michael Atiyah*[15]

The Euler number of the hyperbolic figure eight knot's complement division boundary is 2, since it is isomorphic to a sphere. Its inverse is equal to the self-intersection number (-2), defined by the sum of the external partition parameters squared, divided by their product. And this is equal to the sum of all the 17 fundamental particle charges.

$$\frac{\sum ж_k{}^2}{\prod ж_k} = \left(\frac{ж_1{}^2 + ж_2{}^2 + ж_3{}^2 + ж_4{}^2}{ж_1 \times ж_2 \times ж_3 \times ж_4}\right) = -2$$

Self-intersection number of the external domain

$$\sum_{k=1}^{17} q_k = -2$$

Sum of all 17 fundamental particle charges.

The next question is, how does this connection logic extend? That is, since we know the full internal division balance of a single minimum universe, how do we define the intersection of 2 of these minimum balances, or 3, and so on, can we show that the sequences of compact orientable surfaces and compact non-orientable surfaces, define the combinatorial sequence of stabilization between g-intersecting universes ($\chi = -2 + 2g$)?

Reimann hypothesis

Now that we have a richer way to talk about the Reimann zeta function, we are in a uniquely favorable position for addressing the Reimann hypothesis.

---

[15] 4-dimensional models of nuclei. Sir Michael Atiyah, 2011.

Acknowledgements

I thank George Cassiday for the passion of his ETI class.

I thank Albert Einstein for teaching me true courage of aim, Leonhard Euler for his insightful reach, and Gene Shoemaker (the man on the moon) for his bold example.

I thank my Num for pinky holds, Shangri-La, and for really meaning it.

I thank my $\sqrt{-1}$ Cloud for her delightful imagination.

I thank Elaine, Phil, and Matt Emmi for the way they gave me a key.

I thank Jeff Chapple for all the Alfred stuff, and David Cantu for what may or may not have been said in the cone of silence.

I thank Anaximander, Michael Faraday, Benoit Mandelbrot, Louis de Broglie, David Bohm, Stephen Wolfram, Shelly Goldstein, Detlef Dürr, Nino Zanghí, Grant Sanderson (3Blue1Brown), Derek Muller (Veritasium), Brady Haran (Numberphile), Franck Laloë, Garrett Lisi (PSI), Burkard Polster (Mathologer), Destin Sandlin (Smarter Every Day), Dianna Cowern (Physics Girl), Grigori Volovik, Robert Brady, Ross Anderson, Erwin Madelung, Tim Maudlin, Richard Feynman, Craig Callendar, Christian Wüthrich, Cohl Furey, Yoshio Koide, David Richeson, William Thomson, TED, The Royal Institution, Norman Wildberger (Insights into Mathematics), Michael Atiyah, and Jaigyoung Choe—for their priceless gifts of clarity.

I thank Avi Rubin, Steve Desofi, Johnny Wanda, Nolan Eakins, John Tesla, Nicholas Cecaci, David Humpherys, Jennifer Scharf, and Matthew Fox for their editorial feedback, and Henry Segerman for his beautiful hyperbolic figure eight knot STL artwork.

I thank WolframAlpha.com and Wikipedia.com for their infinite usefulness in the process of discovery. And I thank Steve Desofi, Avi Ruben, David Heggli, and Jerry Gardner for supporting this quest.

Appendix a: graphs

I highly recommend plotting the graphs in this book yourself, zooming in and out, to develop an intuitive sense of the rich fractal nature of these functions. They can be plotted at WolframAlpha.com by entering the following.

Graph 1: plot pi(sinh((1/2*1/(x+iy))^2))^2, {x,-1.5,1.5}

Graph 2: plot pi(sinh((1/2*(x+iy))^2))^2, {x,-6,6}

Graph 3: plot sinh(sinh(1/7*(x+iy)))^(-1)

Graph 4: plot sinh(sinh(1/7*1/(x+iy)))^(-1), {x,-.042,.042}

Graph 5: plot 5/((7)^(1/2)*(3^(1/3)))*(2^10*e^pi)^(-1/8)
gamma(1/4(x+iy))^2/(gamma(1/2*(x+iy)))

Graph 6: plot 5/((7)^(1/2)*(3^(1/3)))*(2^10*e^pi)^(-1/8)
*gamma(1/4*1/(x+iy))^2/(gamma(1/2*1/(x+iy))), {x,-.1,.1}

Graph 7: plot 2pi*5*(cos(7/5*1/(x+iy)))^2, {x,-8,8}

Graph 8: plot 2pi*5*(cos(7/5*(x+iy)))^2, {x,-5,5}

Graph 9: plot (2*cosh(log(7*(x+iy)))*(cosh(5/2*(x+iy)))^2
(cos(7/5(x+iy)))^2), {x,-4,4}

Graph 10: plot (2*cosh(log(7*1/(x+iy)))*(cosh(5/2*1/(x+iy)))^2
*(cos(7/5*1/(x+iy)))^2), {x,-3,3}

Graph 11: plot gamma(x+iy), {x,-8,8}

Graph 12: plot gammaRegularized(1,1/(x+iy))

Graph 13: plot gammaRegularized(1,1/(x+iy)), {x,-.25,.25}

Graph 14: plot gammaRegularized(1,(x+iy))

Graph 15: plot gammaRegularized(1,(x+iy),1/(x+iy))

Graph 16: plot zeta(x+iy), {x,-15,15}

Graph 17: plot zeta(1/(x+iy)), {x,-.4,1.5}

Graph 18: plot zeta(1/(x+iy)), {x,-.2,.2} {y,-.2,.2}

Graph 19: plot zeta(1/(x+iy)), {x,.9,1.1} {y,-.1,.1}

Graph 20: plot pi*(sinh(1/2*1/(x+iy))^2)^2
*[(sinh(sinh(1/7*1/(x+iy))))^(-1)], {x,-5,5}

Appendix b: mass (equations only)

$$m_e = 2V_{fe}\, m_P{}^4 \left(1 + \left(\sinh\left(\sinh\left(\frac{!\,n}{b}\right)\right)^{-1}\right)^{-1} \boxplus\right)$$

$$\left(\frac{m_e}{m_+}\right)\left(\frac{ж_2}{ж_1}\right)^2 = \left(\frac{2}{3}\right)^2 (\alpha_F - 1)^2 \left(1 - \left(\frac{1}{3}\right) e^{3\gamma} \boxplus\right)$$

$$\left(\frac{m_N - m_+}{m_e}\right) = \left(\frac{1}{3}\right)(\mu + 3 + \pi)\left(1 + \left(\frac{2}{3}\right) e^{3\gamma} \boxplus\right)$$

- -

$$\left(\frac{m_H - m_Z}{m_W}\right) = \left(\frac{1}{3}\right)^2 (\mu + 3 + \pi)(2^{-1})\left(1 + \left(\frac{ж_2{}^2}{ж_1{}^3}\right) G_{Ga}{}^{-3} \left(\frac{n}{b}\right) \boxplus\right)$$

$$\left(\frac{m_N - m_+}{m_e}\right) = \left(\frac{1}{3}\right)(\mu + 3 + \pi)(2^0)\left(1 + \left(\frac{2}{3}\right) e^{3\gamma} \boxplus\right)$$

$$\left(\frac{m_d - m_u}{m_e}\right) = \left(\frac{1}{3}\right)(\mu + 3 + \pi)(2^{+1})\left(1 + \left(\frac{ж_2{}^2}{ж_1{}^3}\right) D_{Do}{}^3 \sqrt{\frac{2}{b}} \boxplus\right)$$

external phases

- -

internal folds

$$\left(\frac{m_c - 2m_s}{m_\mu}\right) = \left(\frac{1}{3}\right)(\mu + 3 + \pi)(2^2)\left(1 + 0 \quad \boxplus\right)$$

$$\left(\frac{m_t - m_\tau}{m_b}\right) = \left(\frac{1}{3}\right)(\mu + 3 + \pi)\left(2^{2^2}\right)\left(1 + b\,\frac{\sqrt{Re(\omega_1)}}{ж_1{}^4} \boxplus\right)$$

$$\left(\frac{m_{\nu_\tau} - m_{\nu_\mu}}{m_{\nu_e}}\right) = \left(\frac{1}{3}\right)(\mu + 3 + \pi)\left(2^{2^{2^2}}\right)\left(1 + 0 \quad \boxplus\right)$$

$$\frac{m_b + m_c + m_t}{\left(\sqrt{m_b} + \sqrt{m_c} + \sqrt{m_t}\right)^2} = \left(\frac{2}{3}\right)^{3} (\alpha_F - 1)^{2} \left(1 + \left(\frac{ж_2{}^3}{ж_1{}^2}\right) \frac{V_{fe}{}^4}{2^{n}} \boxplus\right)$$

$$\left(\frac{m_e}{m_+}\right)\left(\frac{ж_2}{ж_1}\right)^2 = \left(\frac{2}{3}\right)^{2} (\alpha_F - 1)^{2} \left(1 - \left(\frac{1}{3}\right) e^{3\gamma} \boxplus\right)$$

$$\frac{m_e + m_\mu + m_\tau}{\left(\sqrt{m_e} + \sqrt{m_\mu} + \sqrt{m_\tau}\right)^2} = \left(\frac{2}{3}\right)^{1} (\alpha_F - 1)^{0} \left(1 - (3\, ж_2{}^2)\ P_{up} \boxplus\right)$$

- -

$$\frac{m_H + m_Z + m_W}{\left(\sqrt{m_H} + \sqrt{m_Z} + \sqrt{m_W}\right)^2} = \left(\frac{2}{3}\right)^{-\frac{1}{2}} (\alpha_F - 1)^{-e^{2\gamma}} \left(1 + \left(\frac{ж_2}{ж_1}\right)^2 \left(2nb\, (V_{fe})^{\frac{1}{3}}\right)^{-\frac{1}{2}} \boxplus\right)$$

$$\frac{m_u + m_s + m_d}{\left(\sqrt{m_u} + \sqrt{m_s} + \sqrt{m_d}\right)^2} = \left(\frac{2}{3}\right)^{0} \gamma\, (\alpha_F - 1)^{0} \left(1 + 0 \boxplus\right)$$

$$\frac{m_{\nu_e} + m_{\nu_\mu} + m_{\nu_\tau}}{\left(\sqrt{m_{\nu_e}} + \sqrt{m_{\nu_\mu}} + \sqrt{m_{\nu_\tau}}\right)^2} = \left(\frac{2}{3}\right)^{-\frac{1}{2}} (\alpha_F - 1)^{+e^{\pi i}} \left(1 - \left(\frac{ж_2}{ж_1}\right)^2 \left(2nb\, (V_{fe})^{\frac{2}{3}}\right)^{-\frac{1}{2}} \boxplus\right)$$

- -

$$\frac{m_e}{m_c}\left(\frac{ж_2{}^2}{ж_1{}^2}\right) = D_{DO}$$

$$\frac{m_e}{m_d} = \frac{\mu}{ж_2{}^2}$$

$$\frac{m_e}{m_u} = \frac{2}{\omega_2{}^{n}}$$

$$\frac{m_e}{m_s}\left(\frac{!n}{\sqrt{n}}\right) = \frac{\sqrt{K_{\_1}}}{ж_2{}^2}$$

$$\frac{m_e}{m_W}\left(\frac{ж_2{}^2}{ж_1{}^3}\right) = \left(Im(\rho_1)\right)^{-3/4}$$

$$\frac{m_e}{m_b}\left(\frac{ж_2{}^2}{ж_1{}^2}\right) = \frac{L_1{}^3}{2n}$$

$$\frac{m_e}{m_t}\left(\frac{ж_2{}^3}{ж_1{}^2}\right) = \frac{L_2}{6n}$$

$$\frac{m_e}{m_H}\left(\frac{ж_2{}^2}{ж_1{}^2}\right) = \frac{b}{2\, j_{0,1}{}^{b}}$$

$$\frac{m_e}{m_Z}\left(\frac{ж_2{}^3}{ж_1{}^2}\right) = \left(\frac{2}{3}\right)^3 L_{LL}{}^{n}$$

$$\frac{m_e}{m_\tau}\left(\frac{ж_2{}^2}{ж_1{}^2}\right) = \left(\frac{3}{V_{fe}{}^2}\right)^2 \left(1 - ж_2{}^2\, sech\left(tan\left(\frac{1}{2}\right)\right) \boxplus\right)$$

$$\frac{m_e}{m_\mu}\left(\frac{1}{ж_1{}^2}\right) = L_{LL}\left(1 + \frac{\boldsymbol{b}\, ж_2}{V_{fe}{}^2}\left(\frac{!\,\boldsymbol{n}}{\sqrt{\boldsymbol{n}}}\right) \boxplus\right)$$

$$\frac{\phi_1 (2\boldsymbol{n})^{-bn}}{m_{\nu_\tau}}\left(\frac{ж_1}{ж_2}\right)^2 = 2^{1/n}\left(1 - (\sqrt{2} - 1)\left(\frac{ж_2}{ж_1}\right)^2 \boxplus\right)$$

$$m_{\nu_e} = 3\,\phi_0 (2\boldsymbol{n})^{-!n}$$

$$m_{\nu_e} = 2\,\phi_0 (2\boldsymbol{n})^{-!n}$$

- -

$$\frac{1}{ж} + ж + \frac{ж^3}{2\pi} = \left(i^i\right)^{-\frac{\pi}{2}} - m_P$$ the hyperbolic vortex equation

$ж_1 = 0.0854245431533310 \ldots$
$ж_2 = 3.667567534854999 \ldots$
$ж_3 = -1.876496039004165(16) + 4.06615262615971(03)i$
$ж_4 = -1.876496039004165(16) - 4.06615262615971(03)i$

$ж_r = 4.47826244916751 \ldots$ hyperbolic vortex radius constant
$ж_\theta = 2.00316562310924 \ldots$ hyperbolic vortex radian constant

$ж_1 ж_2 ж_3 ж_4 = 2\pi$ product

$ж_1 + ж_2 + ж_3 + ж_4 = 0$ complex zero sum

$ж_1 + ж_2 + Re(ж_3) + Re(ж_4) = 0$ real zero sum

$Im(ж_3) + Im(ж_4) = 0$ imaginary zero sum

$ж_1{}^2 + ж_2{}^2 + ж_3{}^2 + ж_4{}^2 = -4\pi$ square sum

$Re(ж_3) = Re(ж_4)$ twin

$Im(ж_3) = -Im(ж_4)$ reflection

$\left(Re(ж_3)\right)^2 + \left(Im(ж_3)\right)^2 = \left(Re(ж_4)\right)^2 + \left(Im(ж_4)\right)^2 = ж_3 ж_4$ pair

$$\left(Re(ж_3)\right)^2 + \left(Im(ж_4)\right)^2 = \left(Re(ж_4)\right)^2 + \left(Im(ж_3)\right)^2 = ж_3ж_4$$ pair

$$\left(\frac{ж_1{}^2 + ж_2{}^2 + ж_3{}^2 + ж_4{}^2}{ж_1 \times ж_2 \times ж_3 \times ж_4}\right) = -2$$ square sum / product

$$ж_1{}^2 + ж_2{}^2 + ж_3{}^2 + ж_4{}^2 = -\left(2\,\Gamma\left(\tfrac{1}{2}\right)\right)^2$$ square sum gamma split

$$ж_r = \sqrt{ж_3ж_4} \qquad ж_\theta = \tan^{-1}\left(\frac{Re(ж_3)}{Im(ж_4)}\right) + \frac{\pi}{2}$$

$$ж_r{}^2 = ж_3ж_4$$

$$\sum_{k=1}^{4} ж_k = 0 \qquad \prod_{k=1}^{4} ж_k = 2\pi$$

$$\sum_{k=1}^{4} ж_k{}^2 = -4\pi \qquad \prod_{k=1}^{4} ж_k{}^2 = 4\pi^2$$

$$-\pi\sum_{k=1}^{4} ж_k{}^2 = \prod_{k=1}^{4} ж_k{}^2$$ hyperbolic vortex partition relation

Appendix c: constants of Nature (equations only)

$$\alpha = ж_1{}^2$$

$$e = ж_1 q_P$$

$$\lambda_C = 2\pi\left(\frac{l_P\, m_P}{m_e}\right)\left(1 + \alpha_F\left(\frac{b}{2}\right)⊞\right)$$

$$K_J = \frac{ж_1}{\pi}\left(\frac{t_P\, q_P}{l_P{}^2\, m_P}\right)\left(1 - \left(\frac{e^{2\gamma}}{Im(\omega_1)^3}\right)\left(\frac{b}{2}\right)⊞\right)$$

$$\hbar = \left(\frac{l_P{}^2\, m_P}{t_P}\right)\left(1 + \sqrt{\frac{b\pi}{\zeta(3)}}\,⊞\right)$$

$$\varepsilon_0 = \frac{1}{4\pi}\left(\frac{t_P{}^2\, q_P{}^2}{l_P{}^3\, m_P}\right)\left(1 - \left(\frac{n^2}{2^n}\right)⊞\right)$$

$$\kappa = \left(\frac{l_P{}^3\, m_P}{t_P{}^2\, q_P{}^2}\right)\left(1 + \left(\frac{n^2}{2^n}\right)⊞\right)$$

$$H_C = \frac{ж_1{}^2}{2\pi}\left(\frac{t_P\, q_P{}^2}{l_P{}^2 m_P}\right)\left(1 - ж_2\, ⊞\right)$$

$$\mu_B = \frac{ж_1}{2}\left(\frac{l_P{}^2\, q_P\, m_P}{t_P\, m_e}\right)\left(1 + \frac{2\pi^2}{ж_2}\,⊞\right)$$

$$\mu_0 = 4\pi\, ⊞\left(1 + ж_2\, e^{\gamma}\sqrt{\frac{3}{2}}\,⊞\right)$$

$$c_1 = 4\pi^2\left(\frac{l_P{}^4\, m_P}{t_P{}^3}\right)\left(1 - ж_2{}^2(\,2\gamma - 1\,)\,⊞\right)$$

$$\sigma_e = \left(\frac{2}{3}\right)4\pi\, ж_1{}^4\left(\frac{l_P\, m_P}{m_e}\right)^2\left(1 + ж_2{}^2(\,L - 1\,)\,⊞\right)$$

$$G_0 = \frac{\text{ж}_1{}^2}{\pi}\left(\frac{t_P\, q_P{}^2}{l_P{}^2 m_P}\right)\left(1 - \frac{\text{ж}_2{}^2}{\mu^3} \boxplus\right)$$

$$a_0 = \frac{1}{\text{ж}_1{}^2}\left(\frac{l_P\, m_P}{m_e}\right)\left(1 + \text{ж}_2{}^2\left(\frac{6}{Im(\rho_1)}\right)^{1/2} \boxplus\right)$$

$$m_u = m_e\left(\frac{\text{ж}_2}{\text{ж}_1}\right)^2 G_{Gi}{}^{-\frac{3}{4}}\left(1 - \frac{sinh\left(csc\left(\frac{4}{3}\boldsymbol{n}\right)\right)}{\text{ж}_r{}^2} \boxplus\right)$$

$$Z_0 = 4\pi\left(\frac{l_P{}^2 m_P}{t_P\, q_P{}^2}\right)\left(1 + \frac{2\boldsymbol{n}^2}{\boldsymbol{b}}\Gamma(x_{min})^4 \boxplus\right)$$

$$\sigma = \frac{\zeta(2)}{2\boldsymbol{n}}\left(\frac{m_P}{t_P{}^3\, T_P{}^4}\right)\left(1 + \left(\frac{P_{up}}{2\,\text{ж}_1}\right)^2 \boxplus\right)$$

$$N_A = \frac{6\,\text{ж}_1{}^2}{e^\gamma}\left(\frac{1}{q_P\, m_P}\right)\left(1 - \left(2\,\text{ж}_2\, P_{up}\right)^2 \boxplus\right)$$

$$R_K = \frac{2\pi}{\text{ж}_1{}^2}\left(\frac{l_P{}^2\, m_P}{t_P\, q_P{}^2}\right)\left(1 + \frac{1}{2}Re\left(i^{i^{\cdots}}\right)\text{ж}_r{}^2 \boxplus\right)$$

$$E_h = \text{ж}_1{}^4\left(\frac{l_P{}^2\, m_e}{t_P{}^2}\right)\left(1 - \text{ж}_2\,\omega_2{}^4\left(\frac{3}{\boldsymbol{n}}\right)^2 \boxplus\right)$$

$$c = \left(\frac{l_P}{t_P}\right)\left(1 - \left(\frac{3}{2}\right) j_{0,1} \boxplus\right)$$

$$c_{1L} = 4\pi\left(\frac{l_P{}^4\, m_P}{t_P{}^3}\right)\left(1 - \sqrt{\frac{3}{2}}\; j_{0,1} \boxplus\right)$$

$$r_e = \text{ж}_1{}^2\left(\frac{l_P\, m_P}{m_e}\right)\left(1 + \boldsymbol{n}\,(\alpha_F - 1)\sqrt{\left(\frac{2}{3}\right)V_{fe}} \boxplus\right)$$

$$\mu_N = \frac{\text{ж}_1}{2}\left(\frac{l_P{}^2\, q_P\, m_P}{t_P\, m_+}\right)\left(1 + \text{ж}_r{}^2\left(\frac{2}{3}\right)\frac{V_{fe}}{n} \boxplus\right)$$

$$c_2 = 2\pi\,(\,l_P\, T_P\,)\left(1 - 2n\, e^{2\gamma}\, C_{CFP}{}^2 \boxplus\right)$$

$$g_\mu = -\frac{1}{4}(\,\alpha_F - 1\,)\sqrt{\frac{V_{fe}{}^b}{n}}\left(1 + \frac{1}{b}\sqrt{\frac{n}{b}} \boxplus\right)$$

$$g_e = -\frac{1}{4}(\,\alpha_F - 1\,)\sqrt{\frac{V_{fe}{}^b}{n}}\left(1 - \left(\frac{\text{ж}_2{}^2}{\text{ж}_1{}^3}\right)\frac{1}{2n^2}\left(\frac{3}{F_{FR}{}^2}\right)^2 \boxplus\right)$$

$$R = \frac{6\,\text{ж}_1{}^2}{e^\gamma}\left(\frac{l_P{}^2}{t_P{}^2\, q_P\, T_P}\right)\left(1 - \left(\frac{\text{ж}_2{}^2}{\text{ж}_1{}^3}\right)\frac{!n}{2n}\left(\frac{3}{F_{FR}{}^3}\right)^3 \boxplus\right)$$

$$\Phi_0 = \left(\frac{\pi}{\text{ж}_1}\right)\left(\frac{l_P{}^2\, m_P}{t_P\, q_P}\right)\left(1 + 8\, G_{Go} \boxplus\right)$$

$$q_c = \pi\left(\frac{l_P{}^2\, m_P}{t_P\, m_e}\right)\left(1 + 8\, C_C \boxplus\right)$$

$$g_+ = \text{ж}_2\,(\,2\, B_1 + 1\,)\left(1 + 3\, C_c{}^{-4} \boxplus\right)$$

$$R_\infty = \frac{\text{ж}_1{}^4}{4\pi}\left(\frac{m_e}{l_P\, m_P}\right)\left(1 - 4\, L_2\, b^{2/3} \boxplus\right)$$

$$\gamma_+ = \frac{\pi}{n} csch^2\left(\frac{1}{2}\right)^2\left(\frac{\text{ж}_2}{\text{ж}_1}\right)\text{ж}_r{}^2\left(\frac{t_P}{q_P\, m_e}\right)\left(1 + \left(\frac{\text{ж}_2}{\text{ж}_1}\right)\frac{1}{G'} \boxplus\right)$$

$$N_\mu = -\left(\frac{\text{ж}_1}{2}\right)\left(\frac{!n}{\Gamma(n) - 1}\right)\left(\frac{l_P{}^2 q_P\, m_P}{t_P\, m_+}\right)$$

$$F = N_A\, \text{ж}_1\, q_P\left(1 + \frac{\text{ж}_r{}^2}{2\pi}\left(\frac{n}{!n}\right) \boxplus\right)$$

$$g_N = -\left(\frac{ж_\theta}{\boldsymbol{n}}\right)\left(3G_{Gi}{}^2\right)^2\left(1 + \Gamma(\boldsymbol{n})\,⊞\right)$$

$$\alpha_G = \left(\frac{m_e}{m_P}\right)^2$$

$$\omega_C = \left(\frac{m_e}{t_P\, m_P}\right)\left(1 - \frac{1}{2}\Gamma(\boldsymbol{n})\,⊞\right)$$

$$S_{mi} = \frac{1}{ж_1}\left(\frac{m_e{}^2}{t_P\, q_P\, m_P}\right)\left(1 - 2\pi\left(3\, Im\left(i^{i^{\cdots}}\right)\right)^2 ⊞\right)$$

$$G = \left(\frac{l_P{}^3}{t_P{}^2\, m_P}\right)\left(1 - \left(\frac{1}{2}\pi\, ж_r{}^2\right)\Gamma(\boldsymbol{n})\,⊞\right)$$

$$k_B = \left(\frac{l_P{}^2\, m_P}{t_P{}^2\, T_P}\right)\left(1 - \left(\frac{1}{2}\pi\, ж_r\right)^2 ⊞\right)$$

$$\frac{r_e}{r_+} = L\sqrt{\frac{3}{2}}$$

$$\frac{em\ force}{strong\ force} = ж_1{}^2$$

$$\frac{r_e}{a_0} = ж_1{}^4$$

$$He^+ = \frac{a_0}{2}$$

$$Li^{2+} = \frac{a_0}{3}$$

$$\psi(r) = \left(\pi a_0{}^3\right)^{-1/2} e^{-r/a_0} = \frac{a_0{}^{3/2}}{\Gamma\left(\frac{1}{2}\right)}\ e^{-r/a_0}$$

Appendix d: geometric identities

geometry	**circumference of a circle**	**area of a circle**
spherical	$2\pi\, sin(r)$	$2\pi(1 - cos(r))$
Euclidean	$2\pi(r)$	πr^2
hyperbolic	$2\pi\, sinh(r)$	$2\pi(cosh(r) - 1)$

geometry	**Pythagorean theorem**
spherical	$cos(a)\, cos(b) = cos(c)$
Euclidean	$a^2 + b^2 = c^2$
Hyperbolic	$cosh(a)\, cosh(b) = cosh(c)$

geometry	**# of parallels**	$\sum \boldsymbol{\theta}$ ***in*** Δ	**cir/diameter**	**curvature**
spherical	∞	$< 180°$	$> \pi$	< 0
Euclidean	1	$180°$	π	0
Hyperbolic	0	$> 180°$	$< \pi$	> 0

fundamental action	**fixed point of fundamental action**	
$cos(x) = x$	D_{Do}	the Dottie number
$coth(x) = x$	C_{CFP}	Real fixed point of the hyperbolic cotangent
$sinh(x) = x$	0	
$sinh^{-1}(x) = x$	0	
$tanh(x) = x$	0	
$tanh^{-1}(x) = x$	0	
$sin(x) = x$	0	
$sin^{-1}(x) = x$	0	

$x^2 + y^2 = 1$ *unit circle*

$cos^2(x) + sin^2(x) = 1$ *unit circle*

$-x^2 + y^2 = -1$ *unit hyperbola*

$-cosh^2(x) + sinh^2(x) = -1$ *unit hyperbola*

$(x^2 + y^2)^2 = -(-x^2 + y^2)$ lemniscate

$(circle)^2 = -hyperbola$ lemniscate

$r^2 = 2a^2 cos(2\theta)$ lemniscate

$$\pm(sin^2(x) + cos^2(x) + sinh^2(x) - cosh^2(x)) = \pm 0$$
$$\pm(sin^2(x) + cos^2(x) - sinh^2(x) + cosh^2(x)) = \pm 2$$

$$\pm(\text{circle} + \text{hyperbola}) = \pm 0$$
$$\pm(\text{circle} - \text{hyperbola}) = \pm 2$$

$$cos(x) = \frac{1}{2}\left(e^{ix} + e^{-ix}\right) \qquad sin(x) = \frac{i}{2}\left(e^{ix} - e^{-ix}\right)$$

$$cosh(x) = \frac{1}{2}(e^x + e^{-x}) \qquad sinh(x) = \frac{1}{2}(e^x - e^{-x})$$

$$cos^2(x) = cosh^2(i\,x) \qquad -e^{-x} = sinh(x) - cosh(x)$$

$$e^{ix} = cos(x) + i\,sin(x) \qquad log(cos(x) + i\,sin(x)) = ix$$

$$e^{\pi i} + 1 = 0 \qquad e^{-\frac{\pi}{2}} = i^i$$

$$e^{xi} = cos(x) + i\,sin(x) \qquad log(-1) = \pi i$$

$$x\,cosh(log\,x) - x\,sinh(log\,x) = 1 \qquad \frac{1}{2}(1 - x^2) + \frac{1}{2}(1 + x^2) = 1$$

$$2\,cosh(log\,x) = \frac{1}{x} + x = e^{log\,x} + e^{-log\,x} \qquad Ei(\,log\,x\,) = li(x)$$

$$\frac{1}{2}\left(\left(\frac{1}{2}(e^{-x} + e^x)\right) + 1\right) = \left(cosh\left(\frac{x}{2}\right)\right)^2 \qquad li(e^x) = Ei(x)$$

$$coth^{-1}(0) = -\frac{\pi}{2}i \qquad coth^{-1}(i) = -\frac{\pi}{4}i$$

$$tan^{-1}\left(\frac{1}{2}\right) + tan^{-1}\left(\frac{1}{3}\right) = \frac{\pi}{4}$$

$$tan^{-1}(x) + tan^{-1}(y) = tan^{-1}\left(\frac{x + y}{1 - xy}\right) \qquad arctan\,addition$$

$$tan^{-1}(\boldsymbol{n}) + tan^{-1}\left(\frac{1}{\boldsymbol{b}}\right) = tan^{-1}(18)$$

the golden ratio = ϕ

$$sinh(\log \phi) = \frac{1}{2}$$

$$\phi = e^{csch^{-1}(2)}$$

$$csch^{-1}(2) = \log \phi$$

$$\phi = 2\,cos\left(\frac{\pi}{n}\right)$$

$$\phi = \frac{1}{2}csc\left(\frac{\pi}{2n}\right)$$

$$\phi = \frac{1}{2}sec\left(\frac{2\pi}{n}\right)$$

$$\log \phi = sinh^{-1}\left(\frac{1}{2}\right)$$

$$\log \phi = csch^{-1}(2)$$

$$\frac{1}{L_{LL}} = sinh(\,C_{CFP}\,)$$

$$\frac{L_{LL}\, e^{\sqrt{L_{LL}{}^2+1}}}{\sqrt{L_{LL}{}^2+1}+1} = 1$$

Where L_{LL} = the Laplace limit and C_{CFP} = the real fixed point of the hyperbolic cotangent

$$\frac{1}{1-x} = x^0 + x^1 + x^2 + x^3 + x^4 + x^5 + \cdots$$

$$\frac{1}{1+x} = x^0 - x^1 + x^2 - x^3 + x^4 - x^5 + \cdots$$

$$\log(1+x) = \frac{x^1}{1} - \frac{x^2}{2} + \frac{x^3}{3} - \frac{x^4}{4} + \frac{x^5}{5} - \cdots$$

$$1^3 + 2^3 + 3^3 + 4^3 + \cdots + k^3 = (1+2+3+4+\cdots+k)^2$$

$$(1+x)^p = \frac{1}{0!} + p\frac{x^1}{1!} + p(p-1)\frac{x^2}{2!} + p(p-1)(p-2)\frac{x^3}{3!} + \cdots$$

$$\frac{\pi}{4} = \frac{1}{1} - \frac{1}{3} + \frac{1}{5} - \frac{1}{7} + \frac{1}{9} - \cdots$$

$$\zeta(2) = \frac{1}{1^2} + \frac{1}{2^2} + \frac{1}{3^2} + \frac{1}{4^2} + \frac{1}{5^2} + \cdots = \frac{\pi^2}{6}$$

$$\tan^{-1}(x) = \frac{x}{1} - \frac{x^3}{3} + \frac{x^5}{5} - \frac{x^7}{7} + \frac{x^9}{9} - \cdots$$

the complex root

$$i = \sqrt{-1} \qquad i^{i^2} = -i \qquad \left(i^{i^2}\right)^2 = -1 \qquad \left(\left(i^{i^2}\right)^2\right)^2 = 1$$

$$i^i = e^{-\frac{\pi}{2}} \qquad \left(i^i\right)^2 = i^{2i} = e^{-\pi} \qquad \left(\left(\left(i^i\right)^2\right)^2\right)^2 = i^{8i} = e^{-4\pi}$$

$$e^\pi = (-1)^{-i} \qquad \sqrt{e^{\pi i}} = i \qquad \left(\left(i^i\right)^2\right)^2 = i^{4i} = e^{-2\pi}$$

$$i^{i^{i\cdots}} = -\frac{W(-\ln(i))}{\ln(i)} = \left(\frac{2}{\pi}i\right)W\left(-\frac{\pi}{2}i\right)$$

$$W(x)\,e^{W(x)} = x$$

$$W(x) = \frac{1}{\pi}Re\int_0^\pi \ln\left(\frac{e^{e^{it}} - xe^{-it}}{e^{e^{it}} - xe^{it}}\right)dt$$

$$\int_0^e W_0(x)\,dx = e - 1$$

$$\int_0^\infty \frac{dx}{(e^x - x^2)^2 + \pi^2} = \frac{1}{1 + W(1)}$$

$$\sum_{k=0}^{n-1} e^{2\pi i\left(\frac{k}{n}\right)} = 0 \qquad n = 2 \text{ gives Euler's identity}$$

$\sum d_k = d_0 + d_1 + d_2 + d_3 + d_4 = 4^2 + 11^2 = 137$ sum of powers

$\sum_{k=0}^{4} \phi_k = \phi_0 + \phi_1 + \phi_2 + \phi_3 + \phi_4$ sum of rotations

$\prod_{k=0}^{4} \phi_k = \phi_0\, \phi_1\, \phi_2\, \phi_3\, \phi_4$ product of rotations

$$\frac{2}{\pi} = \frac{\sqrt{2}}{2} \cdot \frac{\sqrt{2+\sqrt{2}}}{2} \cdot \frac{\sqrt{2+\sqrt{2+\sqrt{2}}}}{2} \cdot \cdots$$

$$\pi = \int_{-1}^{1} \frac{dx}{\sqrt{1-x^2}} \qquad \frac{\pi}{8} = \int_0^1 \sqrt{x(1-x)}\ dx$$

$$\pi = \int_{-\infty}^{\infty} \frac{dx}{1+x^2} \qquad e^{-\zeta'(0)} = \sqrt{2\pi}$$

$$x^t = e^{\log(x)\, t} \qquad 1 = \int_0^e \frac{1}{x}\, dx$$

$\sum_{k=0}^{\infty} (2k)!\, cosh(x) = \sum_{k=0}^{\infty} x^{2k}$ hyperbolic double cover factorization

$$(\log 2)^2 = Li_1(-1)^2 = 4\, coth^{-1}(3)^2$$

$$\log(n!) = \log 1 + \log 2 + \log 3 + \cdots + \log n$$

$b\, \zeta(3) = \zeta\left(3, \frac{1}{2}\right)$ Aprey's split

$$3Re(\omega_1)^4 = \frac{1}{3} Im(\omega_1)^4 = 3\left(\frac{\omega_2}{2}\right)^4$$

$$G_{Go} = \int_0^\infty \frac{e^{-u}}{1+u}\,du \qquad = -eEi(-1) \qquad = \cfrac{1}{2 - \cfrac{1^2}{4 - \cfrac{2^2}{6 - \cfrac{3^2}{8 -}}\cdots}}$$

Where G_{Go} = the Gompertz constant, e = Euler's number, Ei = the elliptic integral, equal to a simple continued fraction.

$$V_h^* - A_h^* = \left(\frac{2}{3}\right)(\mathbf{b} - 2\mathbf{n}) \qquad \text{hypersphere max volume to surface area}$$

Where V_h^* = the dimension at which the n-hypersphere has maximum volume, and A_h^* = the dimension at which the n-hypersphere has maximal surface area.

The partition parameters of the hyperbolic figure eight knot

$n = 5$ number of unique rotations
$b = 7$ break in scale symmetry

$\phi_0 = 5.39125836832313 \ldots + 2\pi i(k) \quad k \in \mathbb{Z}$ 0th external rotation constant
$\phi_1 = 1.61625918175645 \ldots + 2\pi i(k)$ 1st external rotation constant
$\phi_2 = 1.87554596713962 \ldots + 2\pi i(k)$ 2nd external rotation constant
$\phi_3 = 2.17642683817579 \ldots + 2\pi i(k)$ 3rd external rotation constant
$\phi_4 = 1.41678698590795 \ldots + 2\pi i(k)$ 4th external rotation constant

$t_P = 5.39125836832313 \ldots \times 10^{-44}\ s$ Planck time
$l_P = 1.61625918175645 \ldots \times 10^{-35}\ m$ Planck length
$q_P = 1.87554596713962 \ldots \times 10^{-18}\ C$ Planck charge
$m_P = 2.17642683817579 \ldots \times 10^{-8}\ kg$ Planck mass
$T_P = 1.41678698590795 \ldots \times 10^{32}\ K$ Planck temperature

$G_{Gi} = 1.01494160640965 \ldots$ Gieseking's constant
$V_{fe} = 2.02988321281930 \ldots$ figure eight knot complement volume
$e = 2.71828182845904 \ldots$ Euler's number
$\pi = 3.14159265358979 \ldots$ Archimedes' constant
$ж_1 = 0.0854245431533304 \ldots$ 1st hyperbolic vortex partition constant
$ж_2 = 3.66756753485499 \ldots$ 2nd hyperbolic vortex partition constant
$ж_3 = -1.87649603900417 \ldots + 4.06615262615972 \ldots i$ 3rd hvpc
$ж_4 = -1.87649603900417 \ldots - 4.06615262615972 \ldots i$ 4th hvpc
$ж_r = 4.47826244916751 \ldots$ hyperbolic vortex radius constant
$ж_\theta = 2.00316562310924 \ldots$ hyperbolic vortex radian constant
$\alpha_F = 2.50290787509589 \ldots$ alpha Feigenbaum constant
$\delta_F = 4.66920160910299 \ldots$ delta Feigenbaum constant
$\gamma = 0.577215664901532 \ldots$ Euler-Mascheroni constant
$\mu = 1.45136923488338 \ldots$ nontrivial zero of the logarithmic integral
$G_{Ga} = 0.834626841674073 \ldots$ Gauss's constant
$L = 2.622057554292119 \ldots$ lemniscate constant
$L_1 = 1.31102877714605 \ldots$ 1st lemniscate constant
$L_2 = 0.599070117367796 \ldots$ 2nd lemniscate constant
$D_{Do} = 0.739085133215160 \ldots$ Dottie number
$\omega_1 = 0.764977018528596 \ldots + 1.32497062714087 \ldots i$ omega_1
$\omega_2 = 1.529954037057192 \ldots$ omega_2 constant
$P_{up} = 2.29558714939263 \ldots$ universal parabolic constant
$K_{\_1} = 1.745405662407346 \ldots$ Khinchin harmonic mean
$j_{0,1} = 2.40482555769577 \ldots$ 1st root of the Bessel function
$L_{LL} = 0.662743419349181 \ldots$ Laplace limit
$C_{CFP} = 1.19967864025773 \ldots$ Real fixed point of the hyperbolic cotangent

$\rho_1 = 0.5 + 14.1314251417346 \ldots i$ 1st nontrivial zero of the zeta function
$W_{We} = 0.474949379987920 \ldots$ Weierstrass constant
$F_{FR} = 2.80777024202851 \ldots$ Fransén-Robinson constant
$i^{i^{i^{\cdots}}} = 0.438282936727032 \ldots + 0.360592471871385 \ldots i$ i power tower
$m_R = 0.7475979202534114 \ldots$ Rényi's parking constant
$\bar{s}_{lse} = 0.869009055274534 \ldots$ mean line-between-square-edges length
$\sigma_{zs} = 0.311078866704819 \ldots$ Zolotarev-Schur constant
$A_h^* = 7.25694640486057 \ldots$ dimension of maximal n-hypersphere area
$V_h^* = 5.25694640486057 \ldots$ dimension of maximal n-hypersphere volume
$x_{min} = 1.461632144968362 \ldots$ value at which Γ is minimal for + argument
$\Gamma(x_{min}) = 0.885603194410888 \ldots$ minimal value of Γ for + argument
$\zeta(2) = \pi^2/6$ zeta of 2
$\zeta(3) = 1.20205690315959 \ldots$ Aprey's constant
$G_{Go} = 0.596347362323194 \ldots$ Gompertz constant
$\mathcal{F} = 3.359885666243177 \ldots$ reciprocal Fibonacci number
$T_{tet} = 1.927561975482925 \ldots$ tetranacci number
$C_C = 0.643410546288338 \ldots$ Cahen's constant
$B_1 = 0.261497212847642 \ldots$ Merten's constant
$G' = 1.17897974447216 \ldots$ Wilbraham-Gibbs constant
$C_{Murata} = 2.82641999706759 \ldots$ Murata's constant
$G_g = 1.15872847301812 \ldots$ tether length for grazing half the unit circle

17 tessellations of the Planck mass boundary

$m_e = 9.10938370161994 \ldots \times 10^{-31}\ kg$ electron mass
$m_+ = 1.67262192371195 \ldots \times 10^{-27}\ kg$ proton mass
$m_N = 1.67492749802284 \ldots \times 10^{-27}\ kg$ neutron mass
$m_c = 2.27188026398178 \ldots \times 10^{-27}\ kg$ charm quark mass
$m_d = 8.44242715614137 \ldots \times 10^{-30}\ kg$ down quark mass
$m_u = 3.81810683898335 \ldots \times 10^{-30}\ kg$ up quark mass
$m_s = 1.82501207639326 \ldots \times 10^{-28}\ kg$ strange quark mass
$m_b = 7.45149186313980 \ldots \times 10^{-27}\ kg$ beauty (bottom) quark mass
$m_t = 3.08390948667753 \ldots \times 10^{-25}\ kg$ truth (top) quark mass
$m_H = 2.23150010999262 \ldots \times 10^{-25}\ kg$ Higgs boson mass
$m_Z = 1.62556627846185 \ldots \times 10^{-25}\ kg$ Z boson mass
$m_W = 1.43263881046217 \ldots \times 10^{-25}\ kg$ W boson mass
$m_\tau = 3.16754001786349 \ldots \times 10^{-27}\ kg$ tau mass
$m_\mu = 1.88353162775445 \ldots \times 10^{-28}\ kg$ muon mass
$m_{\nu_\tau} = 7.63391385156818 \ldots \times 10^{-39}\ kg$ tau neutrino mass
$m_{\nu_\mu} = 1.61737751049693 \ldots \times 10^{-43}\ kg$ muon neutrino mass
$m_{\nu_e} = 1.07825167366462 \ldots \times 10^{-43}\ kg$ electron neutrino mass

constants of Nature

$\alpha = 7.2973525729552 2 \ldots \times 10^{-3}$ fine-structure constant

$e = 1.602176574059 73 \ldots \times 10^{-19}\ C$ electron charge

$\lambda_C = 2.42631023893941 \ldots \times 10^{-12}\ m$ Compton wavelength

$K_J = 4.83597848400467 \ldots \times 10^{14}\ sC/m^2kg$ Josephson constant

$\hbar = 1.05457172593010 \ldots \times 10^{-34}\ m^2kg/s$ Planck's constant

$\varepsilon_0 = 8.85418781308692 \ldots \times 10^{-12}\ s^2C^2/m^3kg$ electric constant

$\kappa = 8.98755179196986 \ldots \times 10^9\ m^3kg/s^2C^2$ Coulomb's constant

$H_C = 3.87404614816855 \times 10^{-5}\ C^2/m^2kg$ quantized Hall conductance

$\mu_B = 9.27400994938886 \ldots \times 10^{-24}\ m^2C/s$ Bohr magneton

$\mu_0 = 1.25663706143747 \ldots \times 10^{-6}\ mkg/C^2$ magnetic constant

$c_1 = 3.74177185217629 \ldots \times 10^{-16}\ m^4kg/s^3$ 1st radiation constant

$\sigma_e = 6.65246159951664 \ldots \times 10^{-29}\ m^2$ electron Thomson x section

$G_0 = 7.74809172907834 \ldots \times 10^{-5}\ sC^2/m^2kg$ conductance quantum

$a_0 = 5.29177210936601 \ldots \times 10^{-11}\ m$ Bohr electron radius

$m_u = 1.66053906659995 \ldots \times 10^{-27}\ kg$ atomic mass constant

$Z_0 = 3.76730313668332 \ldots \times 10^2\ m^2kg/sC^2$ characteristic impedance

$\sigma = 5.67037441935166 \ldots \times 10^{-8}\ kg/s^3K^4$ Stefan-Boltzmann constant

$N_A = 6.02214076693260 \ldots \times 10^{23}\ 1/mol$ Avogadro constant

$R_K = 2.58128074494007 \ldots \times 10^4\ m^2kg/sC^2$ von Klitzing constant

$E_h = 4.35974472220674 \ldots \times 10^{-18}\ m^2kg/s^2$ Hartree energy

$c = 2.99792458144786 \ldots \times 10^8\ m/s$ speed of light

$c_{1L} = 1.19104286900535 \ldots \times 10^{-16}\ m^4kg/s^3$ spectral radiance

$r_e = 2.81794032269510 \ldots \times 10^{-15}\ m$ classical electron radius

$\mu_N = 5.05078369897026 \ldots \times 10^{-27}\ m^2C/s$ Nuclear magneton

$c_2 = 1.43877687708189 \ldots \times 10^{-2}\ m\,K$ 2nd radiation constant

$g_\mu = -2.00233184179953 \ldots$ muon g-factor

$g_e = -2.00231930436231 \ldots$ electron g-factor

$R = 8.31446261831660 \ldots\ m^2kg/s^2K\ mol$ molar gas constant

$\Phi_0 = 2.06783384804388 \ldots \times 10^{-15}\ m^2kg/sC$ magnetic flux constant

$q_c = 3.63694755171206 \ldots \times 10^{-4}\ m^2/s$ quantum of circulation

$g_+ = +5.58569468933155 \ldots$ proton g-factor

$R_\infty = 1.09737315685870 \ldots \times 10^7\ 1/m$ Rydberg constant

$\gamma_+ = 2.67522187458888 \ldots \times 10^8\ s/kg\ C$ proton gyromagnetic ratio

$N_\mu = -9.66236357094966 \ldots \times 10^{-27}\ m^2C/s$ neutron magnetic moment

$F = 9.64853321242908 \ldots \times 10^4\ C/mol$ Faraday constant

$g_N = -3.82608515049597 \ldots$ neutron g-factor

$\alpha_G = 1.75182147492404 \ldots \times 10^{-45}$ gravitational coupling constant

$\omega_c = 7.76344099445556 \ldots \times 10^{20}\ 1/s$ Compton angular frequency

$S_{mi} = 4.41899541452692 \ldots \times 10^9\ kg/s\ C$ Schwinger magnetic induction

$G = 6.67384038951738 \ldots \times 10^{-11}\ m^3/s^2kg$	gravitational constant
$k_B = 1.38064931695409 \ldots \times 10^{-23}\ m^2kg/s^2K$	Boltzmann constant
$r_+ = 8.77493567970 17 \times 10^{-16}\ m$	proton radius

Where the black digits represent previously known values (either measured or geometrically known), green digits represent extended predictions, and red digits represent discrepancies between prediction and measurement. Note, there are no red digits.

number of measurement digits needing explanation = 439
number of predicted digits verified by measurement = 439
number of additionally predicted digits = 646
total number of prediction digits listed = 1,085

Other books by Thad:

Einstein's Intuition: Visualizing Nature in Eleven Dimensions

Moon Rock: Mare Crisium

Passages

A Perfect Universe

Source Code: the balance of persistence

www.ingramcontent.com/pod-product-compliance
Lightning Source LLC
LaVergne TN
LVHW052252100826
845147LV00001B/22

* 9 7 8 0 9 9 6 3 9 4 2 9 1 *